T0144449

Integrated Flood Risk Management

This book tackles the question of how we can manage flood-related disaster risk such as of floods, landslides, debris flows and storm surges, that have been intensified by increasing exposure, societal vulnerability and climate change and have been generating unprecedented disasters worldwide. It presents recent conceptual developments in disasters, risk and resilience, and surveys UN policies on environment and development as well as disaster management.

Sustainable and resilient development requires an integrated approach and human empowerment. Japan provides a useful example of effective flood management and disaster recovery in its current strategies for river and basin integrated flood management. Very few English-language books offer up-to-date Japanese experience in the context of global trends for students and professionals relevant to a time of climate change and for global application.

Integrated approach requires flood risk management as part of societal development involving all stakeholders of all sectors and disciplines especially in relation to land and water management. The current Japanese exercise in hazard maps, early warning and evacuation, for example, demonstrates the importance of integrated efforts of public-help, mutual help and self-help with scientific knowledge, political will, administrative commitment, news media, community initiatives, etc.

Integrated Flood Risk Management is an ideal textbook for professionals working for public works and environmental protection, hydrologists and engineers, as well as students of disaster management and water resources development.

Kuniyoshi Takeuchi is Professor Emeritus of the University of Yamanashi, Kofu, Japan where he taught hydrology and water resources for 30 years till 2007. He served as the founding director (2006–2014) and later an advisor (2014–2017) of the International Centre for Water Hazard and Risk Management (ICHARM) under the auspices of UNESCO, Tsukuba, Japan. He got his BS (1966), MS (1968) and later DrEng (1982) in civil engineering at the University of Tokyo, and PhD (1972) in city and regional planning at the University of North Carolina at Chapel Hill. He has specialized in surface hydrology, water resource systems and disaster management. His current interest includes, inter alia, transdisciplinary approach for scientific decision-making for building societal resilience to disasters. He served in various professional offices including the chairperson of Inter-Governmental Council of UNESCO IHP for 1998–2000, the president of IAHS in 2001–2005, the chair of IUGG GeoRisk Commission in 2007–2015, a vice-chair of Science Committee of ICSU-ISSC-UNISDR Integrated Research on Disaster Risk (IRDR) in 2009–2015. He is a recipient of several professional awards including IAHS-UNESCO-WMO International Hydrology Prize (2012).

Integrated Flood Risk Management

Basic Concepts and the Japanese Experience

Kuniyoshi Takeuchi

Routledge
Taylor & Francis Group

LONDON AND NEW YORK

Cover image courtesy of Sankei News. Photo taken by EMORI Inui from Sankei News helicopter at 11:37am on 13 October 2019.

A dike breach at Hoyasu of the Chikuma River by the East Japan Typhoon.

First published 2022
by Routledge
4 Park Square, Milton Park, Abingdon, Oxon OX14 4RN

and by Routledge
605 Third Avenue, New York, NY 10158

Routledge is an imprint of the Taylor & Francis Group, an informa business

© 2023 Kuniyoshi Takeuchi

British Library Cataloguing-in-Publication Data
A catalogue record for this book is available from the British Library

Library of Congress Cataloging-in-Publication Data
Names: Takeuchi, Kuniyoshi, author.
Title: Integrated flood risk management : basic concepts and the Japanese experience / Kuniyoshi Takeuchi.
Description: London ; New York : Routledge, 2023. | Includes bibliographical references and index.
Identifiers: LCCN 2022014168 | ISBN 9781032230733 (hardback) | ISBN 9781032230740 (paperback) | ISBN 9781003275541 (ebook)
Subjects: LCSH: Flood damage prevention—Japan. | Flood control—Japan. | Risk management—Japan. | Emergency management—International cooperation.
Classification: LCC TC530 .T35 2023 | DDC 627/.40952—dc23/eng/20220808
LC record available at https://lccn.loc.gov/2022014168

ISBN: 978-1-032-23073-3 (hbk)
ISBN: 978-1-032-23074-0 (pbk)
ISBN: 978-1-003-27554-1 (ebk)

DOI: 10.1201/9781003275541

Contents

Foreword

History is a melody in which the rise and fall of civilizations rhyme with the ebb and flow of rivers. A society's survival and continuance are inextricably linked to how it deals with its water resources, to how it adapts to environmental and societal changes associated with water, and most precisely, to how it handles flood management. Since the dawn of the 20th century, humanity has come to realize that flood management is not simply an engineering pursuit but is also a social undertaking. What needs to be strengthened are not only structural measures such as levees and dikes but also non-structural measures, namely to promote consensus and collaboration among all stakeholders, to educate the citizenry and to build a resilient society with a disaster-preparedness culture.

From a UNESCO point of view, disaster risk reduction (DRR) is a vital pillar in building today's sustainable cities. An integrated approach to flood risk management will help us chart a new course to minimize societal vulnerabilities in the face of natural hazards drastically amplified in an era of irreversible climate change. This work by Professor Takeuchi represents a significant exploration of this integrated approach. It is a textbook for practical use for water professionals, practitioners and managers, and expected to contribute to raising further capacities in flood risk management in Asia and the Pacific region and beyond.

Professor Takeuchi compiled this textbook based on his more than forty years of work in universities and leadership in the UNESCO Intergovernmental Hydrological Programme (IHP) (he was chair from 1998 to 2000, formulating IHP phase VI), and in hydrological and water-related hazard risk management, both globally and in Asia and the Pacific region. Professor Takeuchi was the secretary of IFI (international flood initiative) during his years as Director of ICHARM (International Centre for Water Hazard and Risk Management under the auspices of UNESCO, category 2 center), when he advanced the implementation of theme-1 of IHP VIII.

The two main outstanding points of this textbook are:

1) it is the detailed compilation of background discussion of major themes of the IHP program in its evolution, which is a significant contribution to the IHP program to raise wider understanding.
2) it is a fascinating account of the Japanese experience of flood risk management throughout history. The Asian experience has little been showcased so far and no such compilation has existed until now.

To begin with, this textbook traces the causal links from the outcome of a disaster to the root causes stemming from societal vulnerability. This conceptualization thus directs disaster management efforts to focus more on disaster risk than recovery. It necessitates an integrated management methodology that is centered on human empowerment and social sustainability. Subsequently, based on a rich and informative investigation on the basic concepts and models of disaster and disaster risk, as well as a comprehensive walk through looking into the conceptual evolution of UN policies on the dialectic of environment and development, and then on disaster management, the center piece of this work is a systemic formulation of Integrated Flood Risk Management as a subset of both Integrated Disaster Risk Management and of Integrated Water Resources Management. It thus enables us to understand Integrated Flood Risk Management as a dynamic process in which we aim to obtain the optimal outcome by a balanced twofold effort: to minimize the negative impact of floods by strengthening infrastructure, scaling down vulnerabilities and fostering disaster preparedness, while maximizing the benefits of floods on human welfare and biodiversity for sustainable development.

In the subsequent chapters, we are provided with abundant examples of IFRM drawing from the rich and poignant Japanese experience which bear witness to the centuries' long struggle of a country harnessing all her engineering, social and cultural forces to cope with flood-related disasters. The world has much to learn from this invaluable experience. Most notably, as evidenced by the example of the Ise City's vicennial tradition of renewing its shrine and the populace's excellent performance during flood hazards, special attention should be given to the value of the intangible cultural heritage in building resilience and social capital, in strengthening sustainability of the local community, and in promoting a disaster-preparedness culture, especially during the contemporary development frenzy characterized by urban sprawl.

In his essay, *UNESCO: Its Purpose and its Philosophy*, Sir Julian Huxley as its first Director-General, stated that the organization's mandates in peace, security and human welfare need to be based on some form of humanism. It might not be a coincidence that the philosophy of recovery practiced by the Japanese peasant saint Ninomiya Sontoku reveals the same humanistic concern, who in contemporary eyes is not only a disaster recovery manager, but is also a pioneer of sustainable development. Human empowerment, manifested in a multitude of virtues, values and principles, is our best hope for sustainable development.

Prof Shahbaz Khan, Director and UNESCO Representative to the People's Republic of China, the Democratic People's Republic of Korea, Japan, Mongolia and the Republic of Korea.

Preface

Climate change has been progressing and hydro-meteorological extremes are intensifying. Flood-related hazards including landslides, debris flows and storm surges are thus increasing in both frequency and magnitude in many parts of the world, especially in warm humid regions. At the same time, the world population is increasing and, including in the nations of decreasing population, urban concentration is progressing. In addition, society is changing with economic development, cultural diversity, materialistic advancement, growing interdependency and globalization of which all have been widening the difference between rich and poor, accelerating disaster risk-poverty nexus, making society more vulnerable to natural hazards. Thus flood-related disaster risk has been increasing more than ever. It is an urgent matter for any flood-prone nation to learn and engage in integrated flood risk management by exchanging whatever knowledge, experiences and resources available to human beings.

This textbook is written based on lectures I have been giving mainly at the International Centre for Water Hazard and Risk Management (ICHARM) for students of the National Graduate Institute for Policy Studies (GRIPS) Master Course for Disaster Management since 2007 with the title "Basic Concepts of Integrated Flood Risk Management (IFRM)".

It has been my great pleasure to teach, discuss and learn about IFRM together with students coming from various countries of the world. With a few exceptions from private companies or self-supported, most students are supported by Japan International Cooperation Agency (JICA) and have working experience in disaster management as governmental or semi-governmental officers. Since ICHARM is an engineering-based research institute, the major expertise is engineering-oriented technology such as hydrology, river engineering, sedimentation engineering, remote-sensing and data processing. In addition, management subjects such as integrated flood management, urban flood management and national land management are also given to learn how to organize society to use those technologies and manage disaster risk. Students are interested in both engineering and management technologies.

At the same time, students come to Japan wish to learn how Japan could recover from the complete defeat of the war and manage all sorts of severe natural hazards like typhoons, heavy rains and snows, landslides, debris flows, earthquakes, tsunamis and volcanic eruptions, and achieve a remarkable economic development. The answer to such a question would need a comprehensive study of socio-economic experience of Japan. But such experience must be well reflected in her IFRM, too. This textbook tries to present the introduction of an international trend in IFRM concepts and as its concrete example, the Japanese experience that brought the nation to the current state of flood risk management.

Japan does have advanced technologies and unique experiences in flood risk management to share. Nevertheless, there are few English written textbooks, especially that are written by Japanese and introduce them in the context of international conceptual framework and trend. I hope this textbook, even to a partial extent, serves for such a purpose for students, researchers and any other people interested in IFRM and its Japanese experience. It would be my great pleasure if this book serves as a trigger for them to deepen their understanding and thinking, and an encouragement to face and cope with flood-related disaster risks of their country.

I am deeply indebted to many people for this publication. First, I would like to thank my fellow students, mostly from foreign countries, from whom I learned a lot on their problems, views and experiences and got opportunities to deepen my understanding during lectures at ICHARM/GRIPS/JICA Master and Doctor courses since 2007 and, before, at University of Yamanashi. Their fresh eyes and enthusiastic mind for improvement of their nations were the basis of publishing this textbook. I also owe many colleagues of ICHARM who worked together and shared our vision on the importance of human empowerment for building sustainable society. Professor KOIKE Toshio, the Director of ICHARM, Professor emeritus of the University of Tokyo who kindly reviewed drafts and gave me valuable suggestions and encouragements. Moreover, he arranged ICHARM and Ministry of Land, Infrastructure and Tourism (MLIT) officers to support this publication for checking and finding good data and figures. Professor A.W. Jayawardena, former Research and Training Advisor of ICHARM, now in Honk Kong University, who suggested this publication by showing his own publications and kept encouraging me through reading a draft and giving me precious comments. Professor TANAKA Shigenobu, former Deputy Director of ICHARM, now in the University of Kyoto helped me in various aspects including the introduction of NINOMIYA Kinjiro Sontoku and Inamura no Hi. He provided me with several precious photos he took during site visits with ICHARM students, too. Similarly, ICHARM colleagues Professor OHARA Miho, Dr. IKEDA Tetsuya, Dr. MIYAMOTO Mamoru also reviewed a draft and gave me valuable comments. My special thanks to Prof. Shahbaz Khan, Director of UNESCO Office Beijing, China who kindly provided a warm

Foreword to this book. Also, to ex-ICHARM colleague Dr. SUGIURA Ai, Program Specialist, UNESCO Office Jakarta, Indonesia for her advice on UNESCO and UN-related matters.

In the stage of getting copyright permission of figures and diagrams, I got quite many helps and offers from various organizations. Particularly from MLIT, Water and Disaster Management Bureau, many Regional Development Bureaus and their River or River and Road Offices not only offered permission but also tried to find original figures and their English versions. In particular, I thank Ms. SUDA Yuko, International Office, River Planning Division, Water and Disaster Management Bureau, MLIT to coordinate all those inquiries. My thanks extend to Sumida Hokusai Art Museum, Ise Jingu Shrine, Hiroshima Peace Museum, Inamura no Hi Museum, Minamis-anriku-cho, Komae City, Weather News, Japan Meteorological Agency Kofu Observatory and the Sankei News.

Special thanks to Professor OKADA Norio and Dr. FUJITA Shoichi who kindly redrawn their original diagrams for this publication that they published many years ago. Also, Mr. MOCHIZUKI Seiichi who offered precious photos from his own photo collection.

I hope this book can make all those kind supports useful for the promotion of IFRM in the world.

28 March 2022
Kuniyoshi Takeuchi
Kofu, Japan

Prologue

Before starting the study on basic concepts of integrated flood risk management and the Japanese experience, it is important to understand UNESCO IHP, the UN's leading program on water, and a special relation between Japan and UNESCO.

Establishment of UNESCO and Japanese participation

The predecessor of UNESCO is considered as the International Committee on Intellectual Cooperation (ICIC) established in the League of Nations (former the United Nations) in 1922. In ICIC the twelve eminent scientists participated such as Albert Einstein and Madame Marie Curie. A Japanese scholar, NITOBE Inazo and one of Under-Secretaries General of the League became the founding director of ICIC. In 1926 French Government established the International Institute of Intellectual Cooperation (IIIC) in Paris and carried out the decisions and recommendations of the ICIC. But its activities stopped by the outbreak of World War II (MEXT: website on History of UNESCO).

After the War, preparation to establish UNESCO (United Nations Educational, Scientific and Cultural Organization) was called upon by the governments of the UK and France in London in November 1945 and a charter (UNESCO Charter) was agreed upon saying:

> Since wars begin in the minds of men, it is in the minds of men that the defenses of peace must be constructed.
>
> *(London, 16 November 1945)*

A year later on 4 November 1946, the charter took effect by 20 nations' ratification and the UNESCO was established. The first Director-General was Sir Julian Sorell Huxley of the UK (1946.12–1948.12) and the eighth was MATSUURA Koichiro of Japan (1999.11–2009.11) (MEXT: website on History of UNESCO).

DOI: 10.1201/9781003275541-1

At the time of UNESCO establishment, Japan was under the occupation of the Allied Forces and not accepted to join the United Nations. But after World War II, as Japan was eager to transform the nation to a peaceful country and contribute to the world by promoting peace, she enthusiastically welcomed the initiative of UNESCO. The first voluntary UNESCO Association to support UNESCO activities started from a local level in Sendai City in 1947 well ahead of any other similar activities in other nations. With this enthusiastic support for UNESCO activities, Japan was admitted to UNESCO in June 1951 before San-Francisco Peace Treaty was signed on 8 September 1951. Moreover, the admission of Japan to the United Nations was only in 1956, five years after the admission to UNESCO. It well describes the profound interest of Japan to UNESCO activities. The number of local UNESCO Associations in Japan was about 278 in December 2018 (Japan UNESCO Association: website).

UNESCO Intergovernmental Hydrological Program

There are the following six science programs in UNESCO (UNESCO: website on Natural Sciences Sector; UNESCO: website on MOST):

1. Intergovernmental Hydrological Program (IHP) since 1975, started as International Hydrological Decade (IHD) in 1965.
2. Man and Biosphere Program (MAB) since 1971.
3. International Geoscience and Geopark Program (IGGP) since 2015, started as International Geological Correlation Program (IGCP) in 1972.
4. International Basic Sciences Program (IBSP) since 2005.
5. Intergovernmental Oceanographic Commission (IOC) since 1960.
6. Management of Social Transformations Program (MOST) since 1994.

The original name of IHP was "International Hydrological Program" that was changed to "Intergovernmental Hydrological Program" at the 40th Session of UNESCO General Assembly in 2019. As IOC is a commission independent of UNESCO, the IHP is the oldest science program of UNESCO. The IHP activities origin from UNESCO's International Geophysical Observation Year (IGY) held in 1957–1958. The observation of CO_2 concentration in Mauna Loa, Hawaii started in 1958 on this occasion and the observation of the Antarctic Island also started then. Japanese Antarctic investigation team headed by NAGATA Takeshi opened the Showa base in Ongle Island on 29 January 1957 followed by wintering by the first wintering team headed by NISHIBORI Eizaburo. It was a great challenge for Japan to prepare an icebreaker and other equipment short after the end of the war (Website on "Soya" and Arctic wintering team).

To follow up on the activities of IGY in hydrology, International Hydrological Decade (IHD) 1965–74 started which eventually grew as a permanent program IHP. During the IHD period, quite many important works were conducted for the establishment of the basis of hydrology. Definition of hydrology was one of them. There are two definitions as follows stated in the International Glossary of Hydrology (UNESCO and WMO Panel on Terminology, 1974). The words underlined were added in its 2nd edition in 1992. It was again slightly revised in the 3rd edition in 2012.

(1) Science that deals with the waters <u>above and below the land surfaces</u> of the Earth, their occurrence, circulation and distribution, <u>both in time and space</u>, their <u>biological</u>, chemical and physical properties, their relation ("and their interaction" added in the 3rd edition) with their environment, including their relation to living beings.

(2) Science that deals with the processes governing the depletion and replenishment of the water resources of the land areas of the Earth, and the various phases of the hydrological cycle.

The first theoretical definition was based on the one given in the US Federal Ad Hoc Panel on Hydrology chaired by Walter B. Langbein (Federal Council for Science and Technology, 1962) and the second more practical definition based on Wisler and Brater: Hydrology (1949, 1959).

Other important achievements of IHD include:

• Experimental Basin research: Valdai experimental basin near then Leningrad, USSR (now Sankt Petersburg, Russia) has a long history of observations of many hydrological phenomena. Data available at (Robock et al., 2007).

• World Catalogue of Very Large Floods (UNESCO, 1976). It was succeeded by IAHS, UNESCO and WMO (1984) World Catalogue of Maximum Observed Floods (IAHS Publication 143) and again in 2003 (IAHS Publication 294).

• World Water Balance and Water Resources of the Earth (UNESCO, 1978)

The third publication was a contribution to the IHD program by the USSR National Committee for the IHD chaired by V.I. Korzun where the central role was played by the State Hydrology Institute (SHI), Leningrad, USSR. This book first reported the global-scale assessment of water on the earth such as the total volume of water on the earth is about 1.4 billion km^3, salt water is 97.5% (sea water 96.5%) and fresh water is only 2.5% which consists of permafrost 70%, groundwater 30%, and river, lake and marsh water accessible to human being is only 0.01% of all water on the Earth.

The International Hydrological Program (IHP) started in 1975 succeeding the success of IHD and continuing till present revising the phase every six years as follows except the second phase to adjust for the UNESCO's planning phase.

1975–80 International Hydrological Program (IHP) Phase I
1981–83 Phase II
1984–89 Phase III
1990–95 Phase IV: Hydrology and Water Resources for Sustainable Development
1996–2001 Phase V: Hydrology and Water Resources in a Vulnerable Environment
2002–2007 Phase VI: Water Interactions: Social Challenges at Risk
2008–2013 Phase VII: Water Dependencies, Stresses to Systems and Societal Responses
2014–2021 Phase VIII: Water Security: Responses to Local, Regional and Global Challenges
2022–2029 Phase IX: Science for a Water Secure World in a Changing Environment

Priority Areas of Phase IX:

1. Scientific research and innovation
2. Water Education in the Fourth Industrial Revolution including Sustainability
3. Bridging the data-knowledge gap
4. Integrated water resources management under conditions of global change
5. Water Governance based on science for mitigation, adaptation and resilience

The 9th phase science plan was accepted in the Intergovernmental Council of the IHP held in November 2021. It says that the plan "identifies key water priority areas to support Members States to achieve Agenda 2030 and the Sustainable Development Goals (SDGs), specially water-related SDGs and other water-related global agendas, such as the Paris agreement on climate change, Sendai Framework on Disaster Risk Reduction (DRR) and the New Urban Agenda (NUA)." (UNESCO IHP, 2021).

From the umbrella title of each phase of the IHP, the change in focus from the time of IHD is obvious that the program is evolving from pure science to application of hydrology and water resources for solving societal issues keeping education and risk avert always in.

Note that water security was defined by IHP VIII as follows:

The capacity of a population to safeguard access to adequate quantities of water of acceptable quality for sustaining human and ecosystem health on a watershed basis, and to ensure efficient protection of life and property against water related hazards – floods, landslides, land subsidence and droughts.

(UNESCO IHP, 2012)

UNESCO water centers

In order to contribute for UNESCO activities from member state side, there is a system of UNESCO institutes or centers to be established in and funded by the member state. There are two categories of UNESCO institutes or centers: Category 1 institute or center operates as an integral part of UNESCO directly governed by UNESCO and category 2 institute or center is governed and operated by the member state herself independently but under the auspices of UNESCO. Their mission was identified at the 40th session of the IHP Bureau at UNESCO-IHE, Delft, 13–15 June 2007 as follows:

The overall mission of UNESCO's water-related institutes and centres is to address water security and water-related challenges by regional and global action, through new knowledge, innovative technologies, collaborative interdisciplinary scientific research, networking, training and capacity development, within the framework of the IHP.

There are 29 category 2 water institute or centers under the auspices of UNESCO in the world as of October 2020. They are according to (UNESCO: website on Water-Related Centres):

1. Institute for Water Education (IHE), Delft, the Netherlands since 1957.

   In 1953, the Netherlands was hit by the devastating North Sea flood which broke many dikes and seawalls, killed almost 2000 people and destroyed 4,500 buildings. To prevent such a tragic disaster to happen again, the ambitious Delta Works started. Such experiences became the target of request of training engineers from several developing countries, especially East Pakistan (now Bangladesh) and in 1957 the first training course "International Course in Hydraulic Engineering" started targeting for developing countries. In 1976 the course became "International Institute for Hydraulic and Environmental Engineering (IHE)".

   During 2001–2016, it served as UNESCO-IHE: Institute for Water Education, Category 1 Institute as an integral part of UNESCO. In November 2017, it came back to IHE Delft "Institute for Water Education", a foundation under Dutch law, and operates under the auspices of UNESCO, a UNESCO category 2 institute.

The IHE has been the largest water graduate school offering MS and PhD and more than 23000 engineers graduated from 190 countries. (UN-IHE: website)

2. International Research and Training Centre on Erosion and Sedimentation (IRTCES), Beijing, China established in 1984.
3. International Research and Training Centre on Urban Drainage (IRTCUD), Belgrade, Serbia established in 1987.
4. Regional Humid Tropics Hydrology and Water Resources Centre for South-East Asia and the Pacific (HTC Kuala Lumpur), Kuala Lumpur, Malaysia established in 1999.
5. Regional Centre on Urban Water Management (RCUWM), Tehran, IR of Iran established in 2002.
6. Regional Centre for Training and Water Studies of Arid and Semi-Arid Zones (RCTWS), Sixth of October City, Egypt established in 2002.
7. International Centre on Qanats and Historic Hydraulic Structures (ICQHS), Yazd, IR of Iran established in 2003.
8. Centre for Water Law, Policy and Science, University of Dundee (Dundee Centre), Dundee, United Kingdom established in 2006.
9. International Centre for Water Hazard and Risk Management (ICHARM), Tsukuba, Japan established in 2006.

Its establishment plan was announced jointly by the Director-General of UNESCO and the Minister of MLIT at the Ministerial Conference during the 3rd World Water Forum in Kyoto in March 2003 and three years later it was established on 6 March 2006 hosted by the Public Works Research Institute, the oldest and most distinguished governmental engineering research institute on public works in Japan.

The mission of ICHARM is to serve as the Global Centre of Excellence for water hazard and risk management to help governments and all stakeholders manage risks of water-related hazards at global, national, and community levels. The hazards to be addressed include floods, droughts, landslides, debris flows, tsunamis, storm surges, water contamination, and snow and ice disasters. It has three pillars of activities, that is, innovative research, effective capacity building and efficient information networks, and globally serve as a knowledge hub for best national/local practices and an advisor in policy making. It aims to deliver the best available knowledge and technology to local practices which ICHARM calls the challenge of "localism".

In the capacity development program, in addition to various training programs, jointly with National Graduate Institute for Policy Studies (GRIPS) and scholarship supported by Japan International Cooperation Agency (JICA), ICHARM has a post-graduate education program on water-related disaster management and already 157 MS and 15 PhD students graduated by the end of 2021. It has been also engaged in various research projects on advanced technology and implementation projects on water-related disaster risk assessment and management in various countries in the world. (PWRI: website on ICHARM)

10. European Regional Centre for Ecohydrology (ERCE), Lodz, Poland established in 2006.
11. Water Centre for Arid and Semi-Arid Zones of Latin America and the Caribbean (CAZALAC), La Serena, Chile established in 2006.
12. International Groundwater Resources Assessment Centre (IGRAC), Delft, The Netherlands established in 2007.
13. Regional Centre for Shared Aquifer Resources Management (RCSARM), Tripoli, Libya established in 2008.
14. International Centre on Hydroinformatics for Integrated Water Resources Management, Parque Tecnologico Itaipu Binacional (CHI), Foz do Iguacu, Brazil jointly managed by the Federative Republic of Brazil and the Republic of Paraguay.
15. International Center for Integrated Water Resources Management (ICIWaRM), Alexandria VA, USA established in 2009.
16. Asia-Pacific Centre for Ecohydrology (APCE), Jakarta, Indonesia established in 2009.
17. Centre for the Sustainable Management of Water Resources in the Caribbean Island States (CEHICA), Santo Domingo, Dominican Republic established in 2010.
18. Regional Centre on Groundwater Resources Education, Training and Research in East Africa (RCGRE), Nairobi, Kenya established in 2011.
19. Central Asian Regional Glaciological Centre (CARGC), Almaty, Republic of Kazakhstan established in 2012.
20. Regional Centre for Integrated River Basin Management (RC-IRBM), Kaduna, Nigeria signed in 2013.
21. Centre for Water for Sustainable Development and Adaptation to Climate Change (WSDAC), Pinosava-Belgrade, Serbia established in 2013.
22. International Centre for Water Cooperation (ICWC), Stockholm, Sweden established in 2014.
23. Regional Centre on Capacity Development and Research in Water Harvesting (RCWH), Khartoum, Sudan established in 2014.
24. Regional Centre for Groundwater Management for Latin America and the Caribbean (CeReGas), Montevideo, Uruguay signed in 2014.
25. International Centre for Water Resources and Global Change (ICWRGC), Koblenz, Germany commenced work in 2014.
26. Regional Centre on Water Security (Centro Regional de Seguridad Hídrica) located in Mexico.
27. International Centre for Water Security and Sustainable Management (i-WSSM), Daejeon, Republic of Korea established in 2016.
28. African Regional Centre for Ecohydrology (ARCE), Addis Ababa, Ethiopia established in 2018.
29. Centre on Integrated and Multidisciplinary Water Resources Management (AUTH), Thessaloniki, Greece established in 2018.

REFERENCES

Federal Council for Science and Technology (1962, June) *Scientific Hydrology*. Report of the U.S. Federal Ad Hoc Panel on Hydrology, Council for Science and Technology Chaired by Walter B. Langbein.

IAHS, UNESCO and WMO (1984) *World Catalogue of Maximum Observed Floods*. Prepared by J.A. Rodier and M. Roche. IAHS Publication 143. IAHS Press, Wallingford, UK.

IAHS, UNESCO and WMO (2003) *World Catalogue of Maximum Observed Floods*. Compiled by Reg Herschy. IAHS Publication 294. IAHS Press, Wallingford, UK. http://hydrologie.org/redbooks/a284/RB284.pdf (accessed 15 December 2021).

Japan UNESCO Association. website www.unesco.or.jp/aboutus/ (accessed 10 July 2021).

MEXT: website on History of UNESCO. www.mext.go.jp/en/unesco/title03/detail03/1373237.htm (accessed 11 July 2021).

PWRI: website on ICHARM. www.pwri.go.jp/icharm/index_j.html (accessed 30 October 2020).

Robock, A., K. Vinnikov, C.A. Sclosser and N.A. Speranskaya (2007) *Data set for Valdai Watersheds 1960–1990*. http://climate.envsci.rutgers.edu/soil_moisture/valdai.html (accessed 15 December 2021).

UNESCO (1976) *World Catalogue of Very Large Floods*. UNESCO Press, Paris.

UNESCO (1978) *World Water Balance and Water Resources of the Earth*. USSR IHD Committee Chaired by V.I. Korzun (SHI, Leningrad, 1974). translated version (UNESCO, Paris).

UNESCO: website on MOST. https://en.unesco.org/themes/social-transformations/most (accessed 6 January 2022).

UNESCO: website on Natural Sciences Sector. https://en.unesco.org/themes/science-sustainable-future/about (accessed 6 January 2022).

UNESCO: website on Water Related Institutes and Centres. https://en.unesco.org/themes/water-security/centres (accessed 30 October 2020).

UNESCO IHP (2012) *International Hydrological Programme (IHP) Eighth Phase: Water Security: Responses to Local, Regional and Global Challenges*. Strategic Plan, IHP-VIII (2014–2021). http://unesdoc.unesco.org/images/0021/002180/218061e.pdf (accessed 16 December 2021).

UNESCO IHP (2021) *Draft IHP-IX Strategic Plan: Science for a Water Secure World in a Changing Environment (2022–2029)*. https://en.unesco.org/sites/default/files/ihp-ic-xxiv-ref.1rev_ihp-ix_strategic_plan_in-session_fin.pdf (accessed 16 December 2021).

UNESCO and WMO Panel on Terminology (1974, 1992, 2012) *International Glossary of Hydrology* (WMO Secretariat, Geneva, 1974, 1992, 2012).

UN-IHE: website www.un-ihe.org/institute (accessed 30 October 2020).

Website on "Soya" and Arctic Wintering Team. http://www5a.biglobe.ne.jp/~t-senoo/Ningen/nankyoku/sub_nankyoku.html (accessed 30 October 2020).

Wisler, C.O. and E.F. Brater (1949) *Hydrology*. John Wiley and Sons Inc., New York, 2nd ed., 1959.

Chapter 1

Introduction

There is no such thing as a natural disaster

Disasters triggered by natural hazards have been occurring ever since the appearance of human beings on the earth. But natural hazards are not a disaster but just natural phenomena since if there is no human who receives them as a disaster, there is no disaster. Thus, human beings are the creator of disasters triggered by natural hazards. Why are some natural phenomena hazardous to human beings? Because human beings are vulnerable to the effects of those natural phenomena.

There is no need for an explanation on such an obvious fact. But once disaster management becomes a community issue, such perception is forgotten and becomes a minor view. Human beings exist depending on the gift of nature and community tends to consider the gift always in the range of necessary quality and quantity in time and space. But variation of nature is not bounded in such a range favorable to human beings. Once it exceeds the range, community tends to say "disaster is an act of god," and people cannot resist it. Sometimes this notion is used as an excuse to difficulties that a community failed to avoid.

Such a notion is well reflected in the term "natural disaster" which implies a disaster is caused by nature. It is true that natural hazards are a key to disaster occurrence as there is no disaster if there is no natural hazard. But it is not a satisfactory condition for a disaster to happen. It does not happen even with natural hazards if human society is well settled and not vulnerable to hazards. The word natural disaster may shade this important cause of disaster, the societal vulnerability to hazards.

It is believed that it was the 1755 Lisbon Earthquake in Portugal where the notion of societal vulnerability came into the picture as a cause of earthquake disaster. It was in fact the first occasion that such scientific views on disaster appeared, not only on its cause but also on the recovery. The Great Lisbon Earthquake occurred at 9 am on 1 November 1755 (Kazumori, 2015). It was truly a devastating disaster triggered by an earthquake in magnitude 8.5–9.0 that destroyed about 17,000 out of 20,000 buildings and killed about 70,000 people.

French philosopher Voltaire wrote a long "Poem on the Lisbon disaster" (1755) and put it in his novel "Candide" (1759) saying "it was a sign of God showing His power, glory and might," and expressed his dis-satisfaction to God or doubt to then the divine belief of "everything is good" (optimism) by Gottfried Wilhelm Leibniz. Objecting it, a young French philosopher Jean-Jacques Rousseau sent him a letter saying "nature did not construct the thousands of houses and multi-storey that collapsed" and it's not a matter of God's anger. This letter from Rousseau to Voltaire was considered the beginning of a change in thinking that led to a new way of interpreting disasters, that is, societal vulnerability is the cause of human-induced disasters rather than natural disasters (Matsuda, 2015).

At the same time, German philosopher Immanuel Kant wrote three papers on the cause of earthquake which became the first scientific paper on earthquake although his original view was a support of a wrong concept that seismic activity was caused by heat generated by chemical reaction by iron, sulfur and water underground but initiated the scientific discussion on seismic phenomena.

Since then, geophysical sciences advanced a lot and much became known on natural hazards by now. But about the cause of disaster, it took a long time until societal vulnerability became the central focus. It was only in 1945 when Gilbert White (1945) wrote "Floods are 'acts of God' but flood losses are largely acts of man" in his book "Human adjustment to floods". Finally, in 2000, the UN International Strategy for Disaster Reduction (UNISDR) was established and the following slogan became the UN's leading principle: "There is no such thing as a natural disaster. Only natural hazards." It has been followed by UN Office for Disaster Risk Reduction (UNDRR) renamed from UNISDR in May 2019 (UNDRR: website on Sendai Framework 6th Anniversary).

This is why this textbook avoids the use of the term "natural disaster" but uses only "disaster" or, if necessary, "disaster triggered by natural hazards" following most of contemporary disaster reduction documents such as "At Risk" by Piers Blaikie et al. (1994). It is noteworthy, however, that the expression "natural disasters" is still widely used such as in the UN documents on Millennium Development Goals (MDGs) and Sustainable Development Goals (SDGs). Also, even if it is avoided, it would be impossible not to use such words like flood disasters or water-related disasters. If the word natural disaster is illogical, flood disaster would also be illogical as flood does not necessarily cause a disaster but societal vulnerability does. The important thing is not a word but a concept. This textbook follows the UNISDR's policy to avoid the use of "natural disaster" emphasizing that human society is making a disaster to ourselves and it is necessary to reduce societal vulnerability in disaster management.

From disaster relief to disaster risk management

Another important conceptual shift in disaster management around the last decade of the 20th century was a paradigm shift from emergency relief and response-oriented disaster management to disaster risk management. This shift was strongly led by UN International Decade for Natural Disaster Reduction (UNIDNDR) started in 1990 followed by UN International Strategy for Disaster Reduction (UNISDR) started in 2000 (now UN Disaster Risk Reduction (UNDRR)) that will be discussed in Section 3.2. It is obvious that relief and recovery from catastrophic disaster is much costly than preventing its occurrence in development stage. But most disaster management budget, national or international, has been allocated to relief and response purposes as humanitarian aids have always urgent and high political priority while risk reduction as part of developmental effort is for the future and less urgent.

Dewald van Niekerk (2008) reported the process of how the focus shift was made from disaster relief to disaster risk in UN policy. He described that disaster relief operation had been implemented by various UN agencies, especially UN Development Program (UNDP) established in 1965 and NGOs from their early stages of activities using their development funds. But they realized that most of the developmental funds were allocated to disaster relief and not to development. In order to better coordinate those activities, UN Disaster Relief Office (UNDRO) was established in 1971 (see Section 3.2.2). But its mandate and budgetary basis were weak and could not function as planned and was absorbed to a newly created UN Department of Humanitarian Affairs in 1991 and succeeded by the UN Office of Coordination of Humanitarian Affairs (UNOCHA) in 1992. Currently, the UN provides disaster relief through many agencies, for example, OCHA, World Food Programme (WFP), United Nations Children's Fund (UNICEF), and Food and Agricultural Organisation (FAO).

They first focused on disasters triggered by natural hazards such as East Pakistan/Bangladesh storm surge by Cyclone Bhola in 1970 and Guatemalan earthquake in 1976 but soon after the end of the Cold War in 1989 extended to man-made disasters such as war refugees in Kurdistan, Somalia, and Bosnia in the first half of the 1990s. The relief providers felt that mere provision of immediate necessities of food, water, shelter, and medical care to survivors was not enough and the relief-receiving people or nations also felt that the assistance was necessary to extend to long-term development. In flood-related disasters, this shift was supported by the advancement of forecasting technology which makes preparedness efficient. Thus, Niekerk (2008) concluded that the conceptual shift from relief to risk reduction through development was contributed by "various international disasters, professional constituencies and international organisations" and not a "natural evolution of a discipline".

As for terminology, *relief* is *response* in UNISDR expression and used as *emergency response*. Similarly, *crisis* is *emergency* in UNISDR terminology and *emergency management* includes "preparedness, response and initial recovery steps" (UNISDR, 2009). In this textbook, as discussed in Section 2.4.1, relief (response) operation is considered part of crisis (emergency) management and starts immediately after a disaster happens. When hazards approach and a disaster risk becomes high, the *crisis management* should start to stop it or minimize its impacts by preparedness. Once a disaster happens, human lives and assets in danger should be saved and the expansion of impacts stopped. Then for the affected people, the provision of emergency services starts as relief operation. Survivors from the crisis are in most cases still in high risk and cannot live if no support of food and other essentials is provided. They need help to safely move to recovery step. Therefore within crisis management, salvation from hazards and support for survivors should be seamlessly connected. Similarly, relief operation, recovery process and development as recognized by a number of relief operation cases.

Disaster risk management includes various actions to be taken before a risk turns into a disaster by avoiding, reducing, transferring or accepting risk. On the other hand, disaster management is a word that includes both before, during and after disaster, that is, crisis management, relief operation and disaster risk management. But it should be noted that disaster management is sometimes used in a narrow sense as emergency management. Under integrated disaster risk management (IDRM), all those phases should be well coordinated and integrated and their distinction becomes less strict.

IDRM is a notion of disaster risk management not by a single sector or by sectors working independently of each other but by working together sharing available information and resources. In other words, it should be multi-sectoral and multi-disciplinary approach or transdisciplinary approach including structural and non-structural means that are efficiently coordinated in the total societal system through public, private, community, institutional, administrative, social, economic, engineering, medical and all other means. Although IDRM considers before a disaster happens, as it is necessary to prepare for after a disaster happens and avoid a disaster to repeat, it is basically same as integrated disaster management (IDM). Difference is just a slight difference in emphasis on risk or disaster as a whole.

Global sustainability

Sustainability is the ability of maintaining the intended or expected function of any system. It has long been considered in natural resources management such as groundwater, forests and fish which could be sustained if wisely used but easily depleted if used in a short-sighted consideration. The word sustainability has been much extended in this time of global change and global problems. These include population increase, poverty, hunger, climate change, environmental destruction, decline of biodiversity, disasters, conflicts and diseases. They all are global problems, interconnected with each other and cannot be solved separately. Nobody can be free from them and everybody is a victim and an assailant. How to solve such problems and survive on the Earth environment? It is a global sustainability issue. Our target theme of integrated flood risk management is part of such a global sustainability issue. In order to look into our theme further, it is necessary to set our position in this global perspective.

"Global sustainability" may be defined as:

> The capacity to create and maintain conditions where human and nature can exist in productive harmony fulfilling social, economic and environmental needs of present and future generations.
> *(modified from EPA: website on* What is Sustainability?*)*

In order for human and nature to exist "in productive harmony", for the human side, it is necessary that human well-being is achieved and maintained as it is not the state of sustaining human being if any member is not well in living. Similarly, for nature side, it is necessary for biodiversity to be achieved and maintained as it is not the state of sustaining biodiversity if any member becomes extinct. The actions to achieve such global sustainability are called "sustainable development" which aims to achieve and maintain human well-being and biodiversity for the present as well as the future generations, which will be further discussed later in Section 3.1.3. Global sustainability is the goal to aim at and sustainable development is the methodology to achieve it.

Following this definition of global sustainability, Figure 1.1 is depicted. The circle of human well-being and biodiversity indicates the breakdown of global sustainability into two basic components. In order to achieve this complete sustainability circle, the triangle indicates the necessary and satisfactory conditions to be fulfilled, namely, social sustainability, economic sustainability and environmental sustainability should be achieved. Within this triangle, various mutually interrelated conditions to be secured are indicated, namely, water, air, land, energy, foods, health, business, finance, economy, peace securities, etc. should be satisfied. Once all those securities are satisfied, the attributes of global sustainability are complete, and the state of global sustainability is realized.

But in reality, achievement and maintenance of any of those securities are difficult and always jeopardized by all kinds of risk. Our focus, disaster risk under natural hazards is one of them. Not only disaster risk but also any components of societal system are at risk. Water, air, land, economy, environment, energy, health, above all, peace and culture are subject to risk, and risk reduction needs to be addressed to achieve and maintain global sustainability.

Therefore, the controller of global sustainability is risk reduction, or its wider concept, resilience building. And for risk reduction and resilience building, as Figure 1.1 shows, scientific knowledge and good governance are indispensable and truly the keys. Without the help of science and technology, and good governance, it is totally impossible to reduce risk and build resilience against any attributes of global sustainability. Here, we have to sincerely recognize that both scientific knowledge and governance are human, the acts of human beings. The controller of sustainability is therefore nothing but what and how human being does. It is a solemn reality that such acts only depend on human empowerment. After all, human empowerment is the only sure way for sustainable development to attain global sustainability.

Figure 1.1 Conceptual framework of global sustainability on the basis of human empowerment. The basic attributes of global sustainability are human well-being and biodiversity realized by social, economic and environmental sustainability when all component attributes are secured. But any component of security is subject to risk of insecurity. Such risk should be reduced and resiliency built by science and technology and good governance which we have to realize are supported by human empowerment. After all, human empowerment is the only sure way for sustainable development to attain global sustainability.

An additional point that Figure 1.1 indicates is the reciprocal relation between security and risk reduction. Namely, risk reduction achieves security and, vice versa, security supports risk reduction. It is a positive spiral. At the same time, if risk reduction fails, security fails, which brings another failure of risk reduction and then security. It is a vicious cycle similar to disaster risk–poverty nexus discussed in Section 2.2.1. Human empowerment can only cut such a vicious cycle through science and technology and governance.

Human well-being and biodiversity

About human well-being and biodiversity, this textbook follows the definitions proposed by the Millennium Ecosystem Assessment (MA). It was called for by UN Secretary-General Kofi Annan in 2000 involving more than 1360 experts to assess the consequences of ecosystem change for human well-being and the scientific basis for action needed to enhance the conservation and sustainable use of those systems and their contribution to human well-being.

In its report "Ecosystems and Human Well-being: Synthesis", biodiversity is defined as (MA, 2005):

> the variability among living organisms from all sources including, inter alia, terrestrial, marine and other aquatic ecosystems and the ecological complexes of which they are part; this includes diversity within species, between species and of ecosystems . . . biodiversity forms the foundation of the vast array of ecosystem services that critically contribute to human well-being.

It is important to note that biodiversity provides ecosystem services to human well-being. MA (2005) indicates that ecosystem services have four components: First, to support circulation of nutrients in soil, water, air and other materials on the earth and control primary products of fauna and flora. Second, to provide foods, fresh water, woods, fiber, etc. Third, to regulate climate, weather, hydrology, disease, water quality, etc. by vegetation through evapotranspiration and microbe through material decomposition, and finally, to form cultural bases of aesthetics, spirit, education, recreation, etc. It is important to note that in order for biodiversity to maintain such ecosystem services, each species has population that can sustain to reproduce itself.

On the other hand, MA (2005) assumes human well-being as follows:

> Human well-being is assumed to have multiple constituents, including the basic material for a good life, such as secure and adequate livelihoods, enough food at all times, shelter, clothing, and access to goods; health,

including feeling well and having a healthy physical environment, such as clean air and access to clean water; good social relations, including social cohesion, mutual respect, and the ability to help others and provide for children; security, including secure access to natural and other resources, personal safety, and security from natural and human-made disasters; and freedom of choice and action, including the opportunity to achieve what an individual values doing and being. Freedom of choice and action is influenced by other constituents of well-being (as well as by other factors, notably education) and is also a precondition for achieving other components of well-being, particularly with respect to equity and fairness.

Thus human well-being should satisfy basic material needs, health, good social relations, security and freedom of choice and action. Those basic components seem closely related to the idea of Franklin Delano Roosevelt's the State of the Union address on 6 January 1941. He introduced the basic four freedoms to be aimed at by democracy, that is, freedom of speech, freedom of worship, freedom from want and freedom from fear. Here, freedom of speech and freedom of worship are the basic part of freedom of "choice" in MA, freedom from want is "basic material needs", and freedom from fear is the important part of "safety" which of course includes disaster safety.

It is obvious that all ecosystem services directly or indirectly, jointly or separately support all five components of human well-being. For example, in order to achieve security from natural hazards, vegetation especially forests and water quality would be vital. Even for freedom of choice and action, availability of a good amount of food, good environment and sound cultural basis are necessary.

Some may raise a question that human well-being should include biodiversity as human beings can never be safe nor happy if biodiversity is not sustained. This is especially true if we consider that human being is part of ecosystem and without biodiversity, human beings may not be able to exist. Nevertheless, such a notion is not employed in MA as well as the definition of sustainability. Instead, human well-being and biodiversity are treated rather as independent and separate concepts. Including both is called global sustainability and the goal of sustainable development is to achieve and maintain global sustainability. This is considered reasonable because if biodiversity is part of human well-being, it is human and not bio. It is too much human-centered and puts biodiversity subordinate to human beings.

About this book

This textbook starts from Prologue introducing Japanese special commitment to UNESCO and its IHP activities to which ICHARM and other water centers belong.

In Chapter 1, basic stands of this textbook are introduced. One is that disaster is not a natural phenomenon but societal vulnerability is responsible and it is on the shoulder of human beings to avoid it. Another is that disaster management efforts should be put before disaster risk turns into a disaster rather than after a disaster happens and relief operation takes place to recover from the aftermath. Finally, position of disaster risk reduction in global sustainability is discussed and the importance of science and technology and governance is stressed, which is nothing but human empowerment, which is the key message of this textbook.

In Chapter 2, basic concepts of disaster and disaster risk are introduced that basically follow the discussion of the book "At Risk, Second edition" by Ben Wisner et al. (2004). The concept described there is the UN's leading principle of the contemporary disaster management policy. First, the structure of risk formation and some basic terminologies are introduced followed by a brief introduction of measuring risk and the world disaster risk data bases. Sections 2.2–2.3 introduce the main "At Risk" models of risk management, the Pressure and Release (PAR) model and the ACCESS model. The PAR model serves for analyzing the structure of disaster risk and how it is formed in society that emphasizes the root causes and dynamic pressure behind visible unsafe conditions of society that faces hazards. The ACCESS model focuses on disaster occurrence at a household level which is important for community risk managers to keep in mind. In Section 2.4 the popular disaster management cycle and a hydro-illogical cycle are explained.

Chapter 3 presents the evolution of UN policies starting from debates on environment and development, then disaster management. Section 3.1 discusses policy development from Stockholm to Rio on environment and development. The emergence of the current UN governing concept, *Sustainable Development* proposed by the World Committee on Environment and Management (WCED) was introduced, followed by its implementation efforts including Agenda 21, the Millennium Development Goals (MDGs) and Sustainable Development Goals (SDGs). Section 3.2 focuses on the evolution of disaster management policy from Yokohama to Sendai, which was the shift of focus from natural hazards to societal vulnerability and from disaster relief to risk reduction. It was mentioned that in such conceptual development Japan played an important role in establishing the WCED and hosting a series of World Conferences on Disaster Reduction in Yokohama, Hyogo and Sendai. In the Sendai framework for disaster risk reduction, near numerical targets were set and its significance was discussed.

The main theme of this book, concepts of integrated flood risk management (IFRM) is presented in Chapter 4, and the case examples in Japan in Chapter 5. Here IFRM is introduced as part of integrated disaster risk management (IDRM) as well as integrated water resources management (IWRM). In Section 4.1, IWRM is introduced first, its appearance, definition, concepts and guideline for practice. IWRM was the only methodology proposed by Agenda 21, UN Conference on Environment and Development

(UNCED) in 1992 for sustainable development of freshwater resources but its concrete implementation procedure was not shown in detail. In this section, introducing the basic components to be integrated in IWRM, that is, nature and hydrological cycle, land and water users, socio-economic system and their interactions, the UNESCO guideline "IWRM at river basin level" is presented as the way for practical use in real world. About IFRM, Section 4.2 presents again its basic concept and methodology. For what and how to integrate, the participatory approach of all related stakeholders including vulnerable people is stressed. As some concrete methodologies, optimal system design procedure and cost-allocation methods are presented followed by a guideline of flood risk management by WMO "Associated Program on Flood Management". Flood risk management is a subset of flood management which includes benefit of flood in addition to risk of flood. In fact once "integrated" flood risk management becomes a focus, there should not be a difference between flood risk management and flood management. A seamless consideration of both risk and benefit is necessary under IWRM.

In Chapter 5, Japanese experiences are presented as an example of IFRM together with IWRM. It is emphasized that Japan is located in a hazardous area and has been struggling with flood-related disasters for a long time. Her experience is impossible to be presented in short so that only recent ones are selectively touched, namely, from the Edo Era to World War II, the dark post-war period, the economic growth period and the current situation in Sections 5.2 to 5.5, respectively. In Section 5.5 judicial structure supporting IFRM as part of IWRM is presented which would be useful to grasp comprehensive efforts for IFRM in Japan. It is also emphasized that IFRM has been exercised since the 1970s by promoting basin-wise management introducing storage and infiltration facilities to prevent runoff increase by development in addition to conventional flood management measures. In Section 5.6, Japanese experiences are summarized in short with some remarks. At the end in Section 5.7, the Great East Japan Earthquakes and Tsunami is introduced together with recent recovery status. Also, presented are other tsunami cases and tsunami preparedness culture "Inamura no Hi".

In Chapter 6, current challenges of Japan in relation to IFRM are presented such as climate change, aging society with depopulation, and as a remedy, compact city approach and social capital for mutual help. In Section 6.2, the importance of social capital is discussed introducing an example city of high social capital, Ise City's "Ise shrine's every 20 years rebuilding tradition (*Ise Jingu shikinen sengu*)". Finally, the transdisciplinary approach is advocated in Section 6.3 as a guiding principle for sustainable development. It is considered necessary for scientific knowledge-based decision-making and it is possible only when the decision-making process is transparent. Some case studies on how it is implemented in Japan, the Philippines and Nepal are touched upon.

In Epilogue, a Japanese peasant saint NINOMIYA Kinjiro Sontoku (1787–1856) is introduced as a disaster recovery manager who saved about 600 devastated villages in the period of Tenpo famine in the 1830s, which occurred in the wake of the Tenmei famine in the 1780s about a half-century before. His recovery method Hotoku Shiho would serve as a universal guide in spirit for disaster risk managers to keep in mind that "Human empowerment is the only sure way for sustainable development".

REFERENCES

Blaikie, Piers, Terry Cannon, Ian Davis and Ben Wisner (1994) *At Risk: Natural Hazards, People Vulnerability and Disasters*, 1st ed. Routledge, London and New York. http://doi.org/10.4324/9780203428764.

EPA: website on What is Sustainability? www.epa.gov/sustainability/learn-about-sustainability#what (accessed 26 November 2020).

Kazumori, Tetsuo (2015, March) Lisbon Earthquake that Initiated the Beginning of Modern Age. In: Masuteru Murozaki et al. (eds.), *Lisbon Earthquake and Its Significance of History of Civilization*. Chapter 1. Research Report, Research Investigation Division, Public Foundation: Hyogo Earthquake Disaster Memorial 21st Century Research Organization, Kobe, Japan.

MA (2005) *Ecosystems and Human Well-being: Synthesis*. Millennium Ecosystem Assessment, 2005. Island Press, Washington, DC, 2005.

Matsuda, Yoko (2015, March) Influence of Lisbon Earthquake on Science – Regarding Seismology of Kant. In: Masuteru Murozaki et al. (eds.), *Lisbon Earthquake and Its Significance of History of Civilization*. Chapter 5. Research Report, Research investigation division, Public Foundation: Hyogo Earthquake Disaster Memorial 21st Century Research Organization, Kobe, Japan.

Niekerk, Dewald van (2008) From Disaster Relief to Disaster Risk Reduction: A Consideration of the Evolving International Relief Mechanism. *TD: The Journal for Transdisciplinary Research in Southern Africa*, 4(2), 355–376. http://repository.nwu.ac.za/bitstream/handle/10394/3952/transd_v4_n2_a5.pdf?sequence=1 (accessed 10 January 2022).

UNDRR: website on Sendai Framework 6th Anniversary. www.undrr.org/news/sendai-framework-6th-anniversary-time-recognize-there-no-such-thing-natural-disaster-were (accessed 16 December 2021).

UNISDR (2009) *2009 UNISDR terminology on disaster risk reduction*. UN International Strategy for Disaster Reduction. https://www.undrr.org/publication/2009-unisdr-terminology-disaster-risk-reduction (accessed 26 November 2020).

White, Gilbert Fowler (1945) *Human Adjustment to Floods – A Geographical Approach to the Flood Problem in the United States*, Ph.D. Dissertation. University of Chicago, Department of Geography, Research Paper No. 29. https://sustainabledevelopment.un.org/content/documents/733FutureWeWant.pdf (accessed 20 November 2020).

Wisner, Ben, Piers Blaikie, Terry Cannon and Ian Davis (2004) *At Risk: Natural Hazards, People Vulnerability and Disasters*, 2nd ed. Routledge, London and New York.

Chapter 2

Occurrence of disaster

2.1 WHAT IS DISASTER RISK?

2.1.1 Fishermen at risk

The picture in Figure 2.1 was drawn, carved and printed by KATSUSHIKA Hokusai (1760–1849). It is Ukiyo-e entitled "The hollow of a wave offshore of Kanagawa" and shows big waves and fishing boats with fishermen. This picture was used, although right and left are reversed, as the cover picture

Figure 2.1 Ukiyo-e "The hollow of a wave off-shore of Kanagawa" carved and printed by KATSUSHIKA Hokusai (1760–1849) (Photo: Courtesy of the Sumida Hokusai Art Museum/DNPartcom). Fishermen are at risk but can avoid it to turn to a disaster or mitigate its impact by reducing vulnerability of fishermen and their villages.

DOI: 10.1201/9781003275541-3

of "At Risk, Second Edition" by Ben Wisner et al. (2004). Using this picture, they explained "fishermen at risk" in a brilliant way. This textbook follows their explanation using the same picture.

Fishermen are at risk. They are exposed to the sea of a high wave which will hit fishermen's boat and they are, with high probability, thrown into the sea at any time. In a psychological sense, disaster might have already occurred to some fishermen to some extent. But in a physical sense, it has not yet happened. If big waves do hit fishermen's boats and boats break or fishermen are thrown into the sea and get drowned, a disaster does happen. But if boats are strong enough, or if fishermen's skill is high enough to row and steer boats, and boats do not overturn, a disaster does not happen. Even if boats get smashed into the sea, if all fishermen have life jackets and are strong enough to swim, a disaster does not happen. Therefore, regardless of a hit of big waves, the occurrence of a disaster is not an inevitable matter but depends on the preparedness of fishermen. The occasion could be just an experience of fishermen exposed to a high-risk situation.

If fishermen's home village is well prepared, fishermen could avoid being disposed to such a risk by early warning, sending rescue boats, equipped with safeguard facilities, etc. Even if the boats are smashed into the sea and many fishermen drowned, if the fishermen's community is well prepared for the probable occurrence of such a disaster, the recovery must be smooth and long-term impacts would be limited. Thus, however large a hazard is, a serious disaster may be avoided. It is the capacity of human side that determines whether disaster occurs or not and, if it occurs, the magnitude of the impacts when a society encounters a hazard.

2.1.2 Terminology

To follow are the definitions of selected basic terminology relating to disaster risk after the 2009 UNISDR Terminology (UNISDR, 2009a) and its revision in UNGA 2016 (UNGA, 2016). Although these two versions, denoted original and revised, are little different, both versions are listed here as the difference would highlight some points of discussion:

Disaster

(original) A serious disruption of the functioning of a community or a society involving widespread human, material, economic or environmental losses and impacts, which exceeds the ability of the affected community or society to cope using its own resources.

(revised) A serious disruption of the functioning of a community or a society at any scale due to hazardous events interacting with conditions of exposure, vulnerability and capacity, leading to one or more of the following: human, material, economic and environmental losses and impacts.

In the original version, disaster is characterized by three components: a serious disruption of the functioning of a community or a society; widespread human, material, economic or environmental losses and impacts; and that the consequence exceeds the ability of the affected community or society to cope using its own resources. The target disaster imaged in the original definition was therefore a societal disaster and not a personal one. In any family, if any member is seriously injured or lost by an accident, it is a disaster. But as long as the UN disaster risk reduction activities are concerned, such a personal case is omitted.

In the revised definition, conditions "widespread" and "exceeds to cope using its own resources (needs of the external help)" were omitted, but since the 1st "serious disruption of the functioning of a community or a society" is kept, the essential contents are the same. Under "serious disruption of the functioning", it is obvious that damages in a community or a society must be widespread in their scale and external help is necessary. In the revised one, the cause is specified as "due to hazardous events interacting with conditions of exposure, vulnerability and capacity" which expresses the nature of disaster "there is no such thing as a natural disaster" quite well.

About an individual scale focus of disaster, in this textbook, the method of analyzing the household level disaster impact will be discussed in Section 2.3 the ACCESS model.

There are some other similar words as disaster, which include calamity, catastrophe and black swan events. While calamity and catastrophe indicate characteristics of extraordinary magnitude of a disaster, black swan event is characterized by its rare and unpredictable nature of occurrence that is beyond what is normally expected and its major impacts, after "Odile" in Tchaikovsky's ballet "Swan Lake".

Hazard

(original) A dangerous phenomenon, substance, human activity or condition that may cause loss of life, injury or other health impacts, property damage, loss of livelihoods and services, social and economic disruption, or environmental damage.

(revised) A process, phenomenon or human activity that may cause loss of life, injury or other health impacts, property damage, social and economic disruption or environmental degradation.

There is no substantial change but just a bit of rewording. In this textbook, the concerned are hazards of natural origin mainly from geological, meteorological, hydrological, oceanic sources, or those in combination. In particular, flood-related hazards such as floods, landslides and debris flows are the focus. Such hazards are described by the frequency of occurrence of different intensities for different areas based on historical data and scientific analyses.

As is said in the definition, hazards "may cause" damages. Therefore, it is not a disaster at all. But in quite many languages, there is no term expressing "hazard" but exchangeable with "disaster". Such as floods, landslides, droughts, earthquakes, volcanic eruptions are synonyms to flood disaster, landslides disaster, drought disaster, earthquake disaster, volcanic disaster and so on. They should be carefully distinguished in risk management.

There is a similar word "peril" used instead of "hazard" such as in Peril Classification and Hazard Glossary by IRDR (2014). According to Webster's Seventh New Collegiate Dictionary (A Merriam-Webster, 1967), hazards is "a source of danger" and peril is "exposure to the risk of being injured, destroyed, or lost". This textbook uses "hazard".

Risk

The combination of the probability of an event and its negative consequences.

Disaster risk

(original) The potential disaster losses, in lives, health status, livelihoods, assets and services, which could occur to a particular community or a society over some specified future time period.

(revised) The potential loss of life, injury, or destroyed or damaged assets which could occur to a system, society or a community in a specific period of time, determined probabilistically as a function of hazard, exposure, vulnerability and capacity.

There are no substantial changes but a bit of elaboration. Risk is defined in general as the probability of occurrence of a hazard multiplied by its negative consequences in a specific time and space. Thus, risk is, in a statistical sense, an expected value of damage.

Vulnerability

(original) The characteristics and circumstances of a community, system or asset that make it susceptible to the damaging effects of a hazard.

(revised) The conditions determined by physical, social, economic and environmental factors or processes which increase the susceptibility of an individual, a community, assets or systems to the impacts of hazards.

Exposure

(original) People, property, systems, or other elements present in hazard zones that are thereby subject to potential losses.

(revised) The situation of people, infrastructure, housing, production capacities and other tangible human assets located in hazard-prone areas.

In both vulnerability and exposure, there are no substantial changes but some rewording. The originals are concise.

Coping capacity

(*original*) The ability of people, organizations and systems, using available skills and resources, to face and manage adverse conditions, emergencies or disasters.

(revised) The ability of people, organizations and systems, using available skills and resources, to manage adverse conditions, risk or disasters. The capacity to cope requires continuing awareness, resources and good management, both in normal times as well as during disasters or adverse conditions. Coping capacities contribute to the reduction of disaster risks.

Resilience

(*original*) The ability of a system, community or society exposed to hazards to resist, absorb, accommodate to and recover from the effects of a hazard in a timely and efficient manner, including through the preservation and restoration of its essential basic structures and functions.

(revised) The ability of a system, community or society exposed to hazards to resist, absorb, accommodate, adapt to, transform and recover from the effects of a hazard in a timely and efficient manner, including through the preservation and restoration of its essential basic structures and functions through risk management.

In the revised, "adapt" and "transform" are added which is an elaboration of "the preservation and restoration of its essential basic structure and function". It is true that the damaged system does not have to come back to the original form as long as the essential basic structure and function are kept. That should include transformation to sustainable society. The revised specifies the methodology "through risk management" which is questionable as the methodology of transformation to resilient society may not be restricted to risk management. Many societal activities towards sustainable development such as economic development, community festivals and equity contribute to risk management.

Note that the definition of *resilience* by Intergovernmental Panel on Climate Change (IPCC) Special Report on Extreme Events (SREX) (IPCC, 2012) uses "anticipate", and seems to include "resist" in "absorb" as:

(IPCC SREX) The ability of a system and its component parts to anticipate, absorb, accommodate, or recover from the effects of a hazardous event in a timely and efficient manner, including through ensuring the preservation, restoration, or improvement of its essential basic structures and functions.

It is quite reasonable that IPCC considers the ability of anticipation by scientific prediction is an important attribute of resilience.

2.1.3 Conceptual formula of disaster risk

Disaster risk is often expressed by the following conceptual formula:

$$R = H \times V \tag{2.1}$$

where R is disaster risk, H is natural hazard, V is societal vulnerability and x is a multiplication operator.

It clearly shows that disaster risk R is a function of both hazard H and vulnerability V and not at all hazard H alone. Furthermore, although multiplication sign x is only a conceptual sign, it indicates if there is no hazard, regardless of vulnerability, there is no risk, and conversely if there is no vulnerability, regardless of hazard, there is no risk. In fact, in desert where no people live, there is no disaster; however, severe floods or droughts may occur.

Since the exposure of people or properties in a hazardous area is so definite and easily identifiable part of societal vulnerability, exposure is often separated from vulnerability.

Namely, omitting the multiplicative operator x, the basic formula becomes:

$$R = HEV \tag{2.2}$$

where E stands for exposure of people or property. This is currently the most popular expression of disaster risk.

Another modification is separating coping capacity from vulnerability such as

$$R = HEV_B/C \tag{2.3}$$

where C denotes coping capacity and V_B is basic vulnerability, that is, the remaining vulnerability other than exposure E or coping capacity C. This is especially useful for disaster managers as coping capacity C is what disaster managers or people engaged in disaster management can do and have control for reducing disaster risk while basic vulnerability V_B is out of their direct control. Coping capacity C is a set of disaster risk reduction measures, namely, disaster reduction infrastructure such as dams, levees, seismic-resistant buildings and shelters, and preparedness measures such as observation, early warning, evacuation, building codes, hazard maps, drills and disaster literacy. On the other hand, the basic vulnerability V_B is the basic national or community vulnerability that disaster managers have no control, namely, poor governance, poverty and their resultant state such as low education, poor health and corruption. Those basic conditions of the nation and people are so decisive to determine societal vulnerability to disasters but only a nation as a whole can improve them and disaster managers cannot do much except participating society as a voter or a consumer. This textbook considers this point important as the author considers the majority of readers are already or aiming at engaged in disaster management.

In order to express such conceptual formula more precisely to accord with definitions of disaster risk R, hazard H and vulnerability V, disaster risk R may be expressed as an expected loss in an area over time period T as follows:

$$R = HV = \int_{A,T} p(H)\, L(H) dH \qquad (2.4)$$
$$V = EV_B/C = L(H) \qquad (2.5)$$

where p(H) is probability of occurrence of hazard H, L(H) is loss due to damaging effect of hazard H, A is a target area and T is a specific period of time over which disaster risk is measured. In order to use this formula for the basis of numerical assessment of disaster risk, the loss function L(H) should be identified. This is an extremely difficult task. A bit simpler approach would be selecting A and T in a particular site and time, use the original formula (2.3), variable by variable, for direct calculation. If its multiplication and division in the formula is cumbersome, it may be transformed into a logarithmic form and make the operators only plus and minus such as

$$\ln R = \ln H + \ln E + \ln V_B - \ln C \qquad (2.6)$$

Or simpler modification may be to transform the coping capacity term $1/C$ into $(1-C/C_{MAX})$, then, formula (2.3) becomes:

$$R = HEV_B\,(1-C/C_{MAX}) \qquad (2.7)$$

where C_{MAX} is a hypothetical ultimate coping capacity that may be considered the ideal state of counter-measures that all possible measures are taken. Note that it is impossible to make disaster risk zero by any means. Therefore, C_{MAX} should include, when it is set, the capacity to control uncontrollable remaining risk. This formula is still useful as it is too difficult in practice to numerically identify the level of coping capacity in absolute value and much easier to estimate the rate of the current state of coping capacity to the ideal state.

Practical application of such efforts are available in the World Risk Report published by Bündnis Entwicklung Hilft (Alliance Development Works) in collaboration with the United Nations University Institute for Environment and Human Security in Bonn, Germany since 2011 (BEH, 2011) and now joined by University of Stuttgart (BEH, 2021). The core report of this report is World Risk Index developed by UNU EHS which uses index similar to Equation (2.3) but the basic vulnerability VB is further divided into two components, susceptibility and adaptability.

2.1.4 Disaster databases

Recording disaster loss data is the basis of identifying effective disaster management policy. Rational policy-making is only possible with correct and deep understanding of the reality based on evidence data. Fact-finding, chronical and comparative analyses of disasters, identifying causes and assessing counter-measures are all imperative steps for rational policy-making and possible only with reliable past records, that is, disaster loss data and response records. Nevertheless, collection and keeping of disaster-related data are not an easy matter and not many countries have good disaster loss database that makes rational decision-making very difficult in the nation as well as in global policy-making. It is therefore precious that regardless of such difficulty on data quality and size, there are a few disaster databases in the world.

EM-DAT: The largest and best used worldwide disaster loss database is EM-DAT (Emergency Events Database) at the Centre for Research on the Epidemiology of Disasters (CRED) established in 1973 in Universite Catholique de Louvain – Ecole de Sante Publique, Belgium. It is collaborated with WHO and supported by USAID/OFDA. They have been collecting disaster loss data including country, dates, hazard, cause of hazard, scale, duration, affected areas, affected people, fatalities and economic losses from the official documents of governments, UN and NGOs, insurance companies, and newspapers by rather few dedicated staff. EM-DAT has disaster data on over 22,000 disasters in the world from 1900 to the present (EM-DAT: website).

The criteria of EM-DAT data collection are that one or more conditions in the following are met:

- Ten or more people were reported killed.
- Hundred or more people were reported affected.
- State of emergency was declared.
- International assistance was called for.

EM-DAT data are used for most of statistics compiled and used for policy-making by most of international and UN agencies.

There are a few other global databases, namely:

The Dartmouth Flood Observatory: It was founded in 1993 at Dartmouth College, Hanover, NH USA and moved to the University of Colorado, INSTAAR (Institute of Arctic and Alpine Research) in 2010. Their Global

Active Archive of Large Flood Events includes flood data since 1985 from news, governmental, instrumental, and remote sensing sources, especially image data in vector or raster form by satellites from all over the world. Their archive is called "active" as it includes the current events. The geographic centers of floods in their FloodArchive GIS file from 1985 to 2010 are 3713. *(Dartmouth Flood* Observatory: website on Global Active Archive of Large Flood *Events)*

NatCatSERVICE: Munich-Re GeoRisks Research Unit has been systematically recording disaster loss data from all over the world for decades and has stored them in a natural hazard archive. The database NatCatSERVICE provides information useful for risk assessment. For example, natural catastrophes such as earthquakes and floods can be analyzed separately since 1980 (Munic-Re: website on NatCatSERVICE).

While EM-DAT data are easily accessible to the raw data, the Dartmouth Flood Observatory provides mainly GIS data and NetCatService provides basically the analyzed results. The EM-DAT service is therefore much appreciated by data analysis users.

SHELDUS: SHELDUS (Spatial Hazard Events and Losses Database, US) is a geo-referenced disaster database for the USA which assembled the scattered various aspects of disaster data collected by many different agencies from 1960 to the present. It was originally developed by the Hazards and Vulnerability Research Institute at the University of South Carolina and, since 2018, maintained by the Center for Emergency Management and Homeland Security at Arizona State University (Cutter, 2010; ASU: website).

UNDRR DesInventar Sendai: After UN World Conference on Disaster Risk Reduction in Sendai in 2015, the UNDRR started this database to monitor SDGs and Sendai Framework. More than 89, mainly developing countries contributed the data at the beginning of 2022 (UNDRR: website on DesInventar).

2.2 PAR MODEL

The PAR Model (Pressure and Release Model) and the ACCESS Model are the ones introduced by the book: Piers Blaikie, Terry Cannon, Ian Davis and Ben Wisner "At Risk: Natural Hazards, People Vulnerability and Disasters, First edition" (Blaikie et al., January 1994). This book was published during the United Nations International Decade for Disaster Reduction (UNIDNDR) 1990–1999 and right before the UN World Conference on Natural Disaster Reduction in Yokohama (see Sec. 3.2.1) in May 1994. Since then its view became the theoretical background and the leading principle of UN disaster reduction policy. This textbook follows its second edition, "At Risk, Second edition" published 10 years later by Ben Wisner, Piers Blaikie, Terry Cannon and Ian Davis (Wisner et al., 2004) right before the World Conference on Disaster Reduction in Kobe, Hyogo Prefecture in January 2005.

The core models of "At Risk" proposed for analyzing the causes of a disaster and its remedy are the PAR model and the ACCESS model. The PAR model tries to identify the structure of risk formation in society and possible societal relief from risk while the ACCESS model tries to identify the detailed disaster impact propagation process at an individual household level.

The basic concept of the PAR model is that a disaster occurs when and where societal vulnerability meets with hazards and its magnitude depends on the size of societal vulnerability and the magnitude of hazard. In their terms,

> a disaster occurs at the intersection of two opposing forces: the process generating vulnerability on one side, and the natural hazards (rather abrupt events or sometimes a slowly unfolding natural process) on the other. The image resembles a nutcracker, with increasing "pressure" on people arising from either side – from their vulnerability and from the impact (and severity) of the hazard for those people. The "release" idea is incorporated to conceptualize the reduction of disaster, that is, in order to relieve the pressure, vulnerability has to be reduced.

In short, it says that pressure is created by both sides, vulnerability and hazards, but release of pressure can only be from vulnerability. In this model, hazard is natural phenomena which basically cannot be controlled by human beings and the counter-measures such as dams and levees are considered prevention measures to reduce vulnerability.

Note that "At Risk" book treats structural infrastructure such as dams and levees as a means of reducing natural hazards as shown in their figures of release model. This is misleading against the concept of "flood is an act of god". In order to avoid such an ambiguity, this textbook treats the reduction or mitigation of the impacts of damaging effects of a hazard as an important part of coping capacity, that is, vulnerability reduction rather than hazard reduction.

2.2.1 Pressure model

The basic structure of PAR model is to analyze the detailed composition of societal vulnerability that meets with hazards. Figure 2.2 shows that the progression of vulnerability has three steps: root causes, dynamic pressure and unsafe conditions. The unsafe conditions are the ones with which people face hazards and everybody notice when a disaster occurs such as exposure to risk area, poor infrastructure and unpreparedness. But the PAR model says there are reasons why such conditions are created. They are the

The Progression of Vulnerability

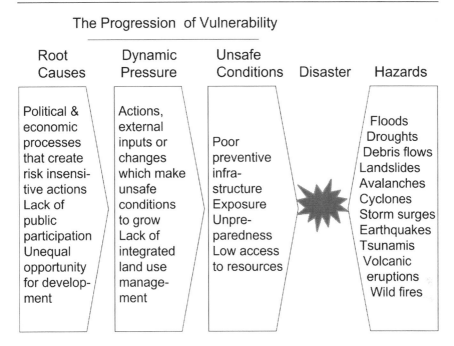

Root Causes — Dynamic Pressure — Unsafe Conditions — Disaster — Hazards

Root Causes	Dynamic Pressure	Unsafe Conditions	Hazards
Political & economic processes that create risk insensitive actions Lack of public participation Unequal opportunity for development	Actions, external inputs or changes which make unsafe conditions to grow Lack of integrated land use management	Poor preventive infrastructure Exposure Unpreparedness Low access to resources	Floods Droughts Debris flows Landslides Avalanches Cyclones Storm surges Earthquakes Tsunamis Volcanic eruptions Wild fires

Figure 2.2 The pressure model to increase pressure between vulnerability and hazards: the progression of vulnerability. Modified from "At Risk" (Wisner et al., 2004).

dynamic pressure that creates unsafe conditions and root causes that are situated even behind, create dynamic pressure and let it realize unsafe conditions. First, the nearest conditions unsafe conditions are examined and then dynamic pressure and finally root causes.

The *unsafe conditions* at risk point that increase the impact of hazards would include such as:

- Population and properties exposed in risk areas, for example, hazardous area, densely populated area and unprotected area.
- Poor physical infrastructure against extreme hazards such as dams, levees, sea walls, check dams, slope protection and seismic-resistant buildings.
- Lack of infiltration and storage facilities and urban development control.
- Weak houses and buildings and lack of building codes enforcement.
- Poor early warning and evacuation system.
- Poor observation system, poor or lack of availability of hazard maps and their use.
- Lack of community preparedness, education, training, drills, disaster

literacy, etc.
- Weak emergency response and rescue system, poor recovery assistance program.

"At Risk" explains unsafe conditions from a more institutional and societal basis and focuses on poor and weak people as follows:

> They (*unsafe conditions*) include such as hazardous location, unable to afford safe buildings, lacking effective protection by the state, engaged in dangerous livelihoods, having minimal food entitlements, initial level of well-being of the people, access to tangible (cash, shelter, food stocks, agricultural equipment) and intangible resources (networks of support, knowledge for survival and sources of assistance, morale and ability to function in a crisis).

Those factors seem more indicative of the conditions in poor governance countries. In most nations, the typical unsafe conditions would relate to the availability of protective infrastructure and preparedness.

Most items of unsafe conditions fall in either exposure or lack of coping capacity and may be improved by disaster management efforts. But the PAR model does not stop here but asks, "Why and how those conditions appear?" There are clear reasons that make society stay in those conditions. They are what "At Risk" calls dynamic pressure, that is, some particular processes and activities that create those unsafe conditions, such as lack of institutional arrangements and some external forces. In fact, the dynamic pressure translates or channels the further deep root causes into unsafe conditions.

The *dynamic pressure* includes such processes and activities as:

- Population increase, rapid urbanization, urban concentration, depopulation and aging and other demographic processes.
- Poor migratory/settlement control into risk area.
- Industrialization, deforestation, hydro-power generation, mining, groundwater pumping, and other especially unregulated development.
- Lack of institutional arrangements for coordination among sectors, disciplines, governments and communities of various levels for disaster management.
- Lack of laws and regulations, projects enactments and lack of law enforcement.
- Poor investment for infrastructure for disaster prevention and mitigation, science and technology, education, observation and forecasting.
- Lack of integrated land and water resources management.
- Conflicts, disasters, poor recovery processes, epidemics, economic crisis, foreign debt, export promotion, etc.

"At Risk" says:

> Dynamic pressures are processes and activities that translate the effects of root causes both temporally and spatially into unsafe conditions or channel the root causes into particular form of unsafe conditions. (Capitalism and neo-liberalism are root causes.) They would include such as epidemic disease, rapid urbanization, wars and conflicts, foreign debt, some structural adjustment programs, export promotion, mining, hydro-electric power development, deforestation, rural-urban migration etc.
>
> It is important to note that these pressures are not necessarily bad and vulnerability-inducing per se. There is a tendency in neo-populist and radical development writing to damn these pressures indiscriminately, without examining their particular historical and spatial specifications. In short, PAR needs thorough research that is locally and historically based. It necessitates micro-mapping of unsafe conditions.

Indeed, some of those such as industrialization, hydro-power generation and development are not at all bad actions but balance with care in disaster risk is important. But such imbalance is often brought by more profound reasons behind dynamic pressure, that is, root causes that tend to create those imbalance processes and activities preventing people or governments to take proper disaster risk reduction arrangements.

The *root causes* are in economic, political & socio-cultural conditions such as:

- Inequality in economic opportunities (poverty, separation between rich and poor, disaster risk-poverty nexus, inefficient use of international aids).
- Lack of political will (imbalance in priority setting between economic opportunities and disaster risk reduction).
- Inequity in political voices (lack of participatory approach, little voices of the poor, disaster-affected people and other weak people, strong voices of the rich, gender inequality).
- Lack of endogenous capacity (low behavioral discipline, corruption, lack of leadership, disaster illiteracy).

"At Risk" says:

> Root causes are most distant from the disaster reality on the spot and often invisible and taken for granted. The most important root causes are economic, demographic and political processes. These affect the allocation and distribution of resources, among different groups of people.

They emphasize that imbalances in economic opportunities and political power are mutually reinforcing and that the people economically marginal tend to be of marginal importance to those who hold economic and political power and are left in a vicious cycle of insecure livelihood, low priority in government support and loss of their own confidence. They are not only cyclic but also often mutually reinforcing and forms the basis of a disaster risk-poverty nexus (UNISDR, 2009b). Note that the impact of such imbalances does not stay in some groups of people but extends to a whole nation. It weakens the total economic power of the nation and keeps the nation poor by preventing necessary investment for disaster risk reduction to make it possible to accumulate wealth.

2.2.2 Release model

Now how to stop the progression of vulnerability and release the pressure of vulnerability balloon? That is the release model. There is no way other than reversing progression of vulnerability by regression of vulnerability, that is, progression of safety as Figure 2.3 shows. In order to reverse the direction of the progression of vulnerability, vulnerability pressures of all balloons should be released which is difficult even from a single balloon. About natural hazards, it is impossible to reduce the pressures (or magnitude) except

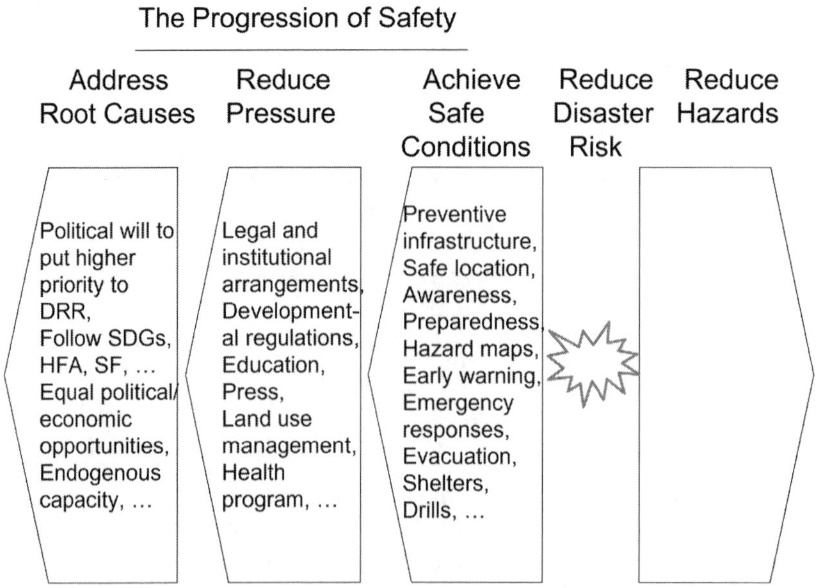

The Progression of Safety

Address Root Causes	Reduce Pressure	Achieve Safe Conditions	Reduce Disaster Risk	Reduce Hazards
Political will to put higher priority to DRR, Follow SDGs, HFA, SF, ... Equal political/ economic opportunities, Endogenous capacity, ...	Legal and institutional arrangements, Development-al regulations, Education, Press, Land use management, Health program, ...	Preventive infrastructure, Safe location, Awareness, Preparedness, Hazard maps, Early warning, Emergency responses, Evacuation, Shelters, Drills, ...		

Figure 2.3 The release model to reduce pressure of vulnerability: progression of safety. Modified from "At Risk" (Wisner et al., 2004).

for a few hazards created by human activities such as landslides by carefully constructing roads in mountain slopes and dam breaks by advanced engineering design and construction works.

The most difficult would be root causes as they are mostly a governance issue. To reverse their direction would be more difficult if they are related to the basic structure of the nation reflecting the history of the nation including colonialism and ideology. But some causes such as political will or priority setting may not be so difficult if the nation follows and implements the international agreements like Sustainable Developmental Goals or Hyogo Framework for Action or Sendai Framework for Disaster Risk Reduction (see Section 3.2). For such political priority setting, various professional conferences have been held. For them, as "At Risk" mentions, there are pessimistic views such as "conferences and the airing of statements of concern, declarations, objectives and principles therein are simply a waste of time". But such views are incorrect. Those meetings draw attention of decision-makers and influence their actions. Quite often, those meetings are held with the support of governments and, still more, in the case of UN initiatives, declarations do not pass without member countries' agreement. The UN agreements are the government committed policy and there is no reason for government leaders to officially deny them.

"At Risk" further writes that some say "corrupt elites in some nations have a self-interest in continuing the cycle of disaster and reconstruction of the status quo because they benefit financially and politically from the flows of foreign relief." These pessimistic views are to some extent undeniable but should be overcome by the more efforts of exchange of information, new initiatives, improved procedures and patience with commitment of time and efforts. It would take time but gradually improve. In fact, it is quite often the case that young diligent students, researchers or scholars eventually become such elites. It is not necessarily a matter of somebody else but all individuals have the potential to become so. Such potential can be avoided by communication with others sharing common concerns, exchange of enlightened ideas, especially international advocacies. It is important to note that "Individuals can make difference". If there is a will, one is not alone. Pessimistic views tend to prevent improvements.

About the dynamic pressures and unsafe conditions, there may be many tractable possibilities to release vulnerability pressure by efforts of disaster risk managers and their offices. Potential improvement may be in people's nearby daily life such as risk consciousness, awareness raising among residents and better preparedness by community members. Such self-consciousness may support disaster managers to promote improvements and additional budgeting for infrastructure, preparation and use of hazard maps, formation of community networks for mutual help, to stop squatters to settle in risk areas, build stronger houses and many other things in everyday community life.

Such awareness of residents would encourage higher-level officers to follow Hyogo and Sendai Frameworks and make better coordination, higher priority setting to risk reduction in developmental decision-making and better exercise of integrated disaster risk management.

As a diagnosis of Release Model, it cannot be over-optimistic and should anticipate many obstacles to face and overcome. But it is true that all the political efforts origin from people's mind, awareness and consciousness and quite much influenced by global trends such as of the UN's initiatives. It is a process of a combination of top-down and bottom-up efforts. Realization of societal change is only by people's empowerment. No external leadership alone can realize the change.

The following section introduces the "At Risk" views on how to release vulnerability pressure which would serve as a guide for using the Release Model.

2.2.3 Risk reduction objectives: CARDIAC (heart)

"At Risk" proposes seven disaster risk reduction objectives to release vulnerability pressure, which is what they call CARDIAC, which literally means "heart". They are the following:

1. **Communicate** understanding of vulnerability: Understand and communicate the nature of hazards, vulnerabilities and capacities.
2. **Analyse** vulnerability: Conduct risk assessment by analyzing hazards, vulnerabilities and capacities.
3. Focus on **Reverse of PAR** model: Reduce risk by addressing root causes, dynamic pressures and unsafe conditions.
4. Emphasize sustainable **development**: Build risk reduction into sustainable development.
5. **Improve livelihood**: Reduce risk by improving livelihood opportunities,
6. **Add recovery**: Build risk reduction into disaster recovery.
7. Extend to **culture**: Build a safety culture.

It is clear that they all well accord with the priorities agreed and declared by UN member countries in the UN World Conferences on Disaster Reduction as Hyogo Framework for Action (HFA) in 2005 and Sendai Framework for Disaster Risk Reduction (SF) in 2015 as will be extensively discussed in Section 3.2, namely:

"**Communicate** understanding vulnerability" and "**Analyzing** vulnerability" are in the 2nd priority of HFA "identify, assess and monitor disaster risks" and the 1st priority of SF "understanding disaster risk". "**Reverse of PAR** model" and "sustainable **development**" are in the 4th priority of HFA "reduce underlying risk". "**Add recovery**" is in the 4th priority of SF "building back better". "Extend to **culture**" is the 3rd priority of HFA "build a culture of safety and resilience at all levels". "**Improve livelihood**" is mentioned a number of times in both HFA and SF as the one to be achieved and protected.

CARDIAC truly indicates the heart of disaster risk reduction and serves as a guide for releasing vulnerability pressure which is well reflected in the priorities declared in the UN World Conferences on Disaster Reduction. After all CARDIAC is the Release Model of "At Risk" and its conclusion.

2.3 ACCESS MODEL

The ACCESS model focuses on the detail process of disaster occurrence at the pressure point where societal vulnerability meets with hazards and a disaster occurs. It magnifies the point of disaster occurrent closely and see how a disaster develops at household level with a different composition of family members and a different relation within a community's power structure. Such a view is important for disaster managers to plan a family support program considering that there are many different families operating under different conditions.

At the pressure point of interactions between natural hazards and households, the ACCESS model tries to identify where and when the disaster starts to develop and when exactly a normal life turns into abnormal. By identifying this, it also makes such questions clear as to why wealthier people often suffer less, and women and children do more than or different from men and adults.

In order to do such an analysis, what is required is detailed accounts of "normal life" before disaster and those of "abnormal life" after disaster. The basic scale that the ACCESS model chose for such account is "the amount of 'access' that people have in the capabilities, assets and livelihood opportunities that will enable them (or not) to conduct their life and reduce their vulnerability to a disaster".

The ACCESS model considers a disaster impact as a negative change of people's access to various resources such as information, food, utility, income, transportation, housing, human relation, and any other properties and livelihood. In this model, a sad feeling by loss or injury of family members is considered a health issue but not a direct disaster loss to impact households. Rather, a disaster impact on economic and livelihood potential is considered as a real count of disaster. In other words, in the ACCESS model, mere sadness is not the core of disaster but concrete loss of materialistic and physical conditions of life is the indicator of real disaster.

"At Risk" explains this process by introducing the terminology "access profile" and "access qualification" such as:

- "Normal life does not necessarily turn into abnormal only by meeting with hazards and human or property losses but does when access to resources or access profile of community changes to low profile and access qualification degrades.
- Real disasters happen when access profile drastically changes and the livelihoods or income opportunities do not continue as before.

- Access profile is a collective profile of access to all resources that each individual or household possesses.
- Access qualification is a set of resources and social attributes which are required in order to take up an income opportunity".

Further, it introduces the notion of social relation and domination structure that greatly influence the determination of the access profiles of any household:

1) "Social relations: the flows of goods, money and surplus between different actors (e.g., merchants, urban rentiers, capitalist producers of food, rural and urban households.)
2) Structure of domination: the politics of relations between people at different levels.

They are decided by the characteristics and capacities of individuals. Class in society, education levels, income levels, gender, ages, health etc. make much difference".

The ACCESS process is described in Figure 2.4.

A specific hazard (3) occurs in a specific time and location with a specific magnitude and form (speed, mode, contamination, debris, etc.) (4) and hit community (5). A specific unsafe condition such as disposure in a risk area (2) determines the specific household livelihood protection level of normal life (t1-tn) according to its position in social relations (1a) and structure of domination (1b) in a society and its aggregation determines the social protection level of a community (1). Such community (1) meets with the triggering event (5) which affects not only households but also their social relations and structure of domination in the community as any disaster impacts on community members and assets immediately change their relative position in the social relations and in structure of domination. In such a social position and given social protection level, each household meets with a triggering event and starts the transition to disaster (6).

The detailed process of transition to disaster (6) is depicted in Figure 2.5 where people's access to safety determines the level of transition to disaster of each household. Namely, each household gets affected according to its family member's contingent situation such as who, where and when meets with the triggering event occurred and what was his/her position in social relations and structure of domination. Such family members' situation determines the resultant household access profile and access qualification for income opportunities, which determines the pay-off (income) and, depending upon household and community coping mechanisms, the form of abnormal life of each household.

This is the first-round impact on normal life and it is not the end. Disaster develops in a process (7) depending on various reactions, coping and

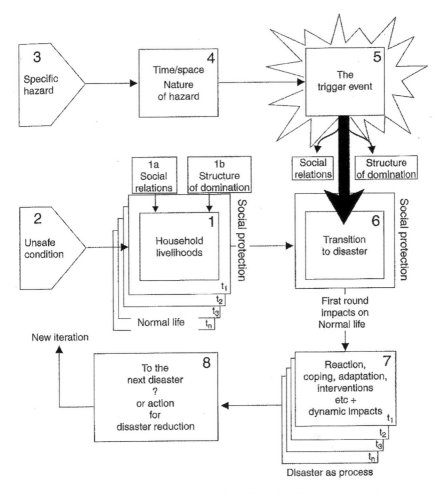

Figure 2.4 The ACCESS model in outline. After "At Risk" (Wisner et al., 2004). Household under unsafe conditions transits to a disaster hit by a hazard according to societal protection. Its impacts and risk reduction actions determine the household against the next disaster.

adaptation actions and public or private interventions. The reactions to this disaster decide the prevention, mitigation and preparedness actions (8) and go to the next round of meeting another hazard. This suggests a cycle that the impacts of a disaster form the condition of a household for another hazard. Since it is difficult to improve the household condition, the cycle tends to become a vicious cycle which is often referred as a disaster risk-poverty nexus as mentioned in the UNISDR GAR report (UNISDR, 2009b). The PAR and the ACCESS models would help finding a way to prevent falling into the risk-poverty nexus and, if fallen, step out and recover from it.

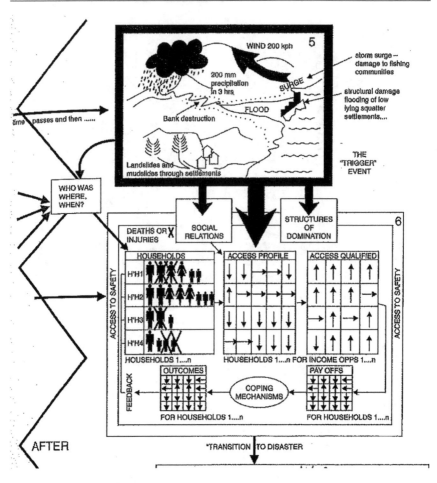

Figure 2.5 The ACCESS model in detail on steps the triggering event and the transition to disaster. After "At Risk" (Wisner et al., 2004). Different households receive different impacts according to who, where and when family members meet with hazard under what societal relation.

2.4 DISASTER MANAGEMENT CYCLE AND HYDRO-ILLOGICAL CYCLE

2.4.1 Disaster management cycle

Necessary actions in disaster management are often classified into the following four phases: Preparedness, Emergency response, Recovery, and Prevention/mitigation as shown in Figure 2.6.

This management process is called a disaster management cycle. The activities involved in each phase are listed in the following. Table 2.1 shows

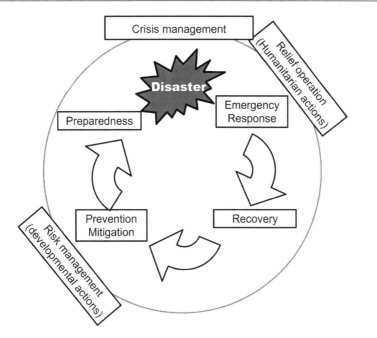

Figure 2.6 Disaster management cycle. While risk management covers all phases, especially preparedness, recovery and developmental actions, crisis management covers mainly emergency response and relief operation for humanitarian actions after a disaster happens.

some concrete examples that the Asian Disaster Reduction Center (ADRC) compiled with slight modifications to make it consistent to Figure 2.6.

1. Preparedness: Preparedness is a control of how risk turns into a disaster. A good preparation can make a disaster small and recover fast. It starts from education, research, awareness raising and extends to observation networks, early warning system, shelter construction, evacuation plan, rescue operation plan, hazard mapping, community response plan, training exercises and drills, etc.

   When hazards approach, preparation is necessary for officers in charge and voluntary helpers to be ready for patrol, emergency operation and victims rescue, to make sure facilities such as sandbags, rubber boats, ropes, search lights, speakers, batteries and first-aid supplies available for dike protection, stopping inundation, rescuing victims, supporting evacuees, etc. and make sure infrastructure and facilities such as pumps, gates, emergency electric generators, etc. operational.

   When hazards come and early warning to be issued, data collection and processing, analyzing, forecasting, information dissemination and issuance of administrative instruction are necessary for which well thought and precise instruction messages should be prepared.

2. Emergency response: When hazards approach and a considerable risk is anticipated, it is already an emergency stage. Early warning should be issued and make sure it is disseminated. Patrols and evacuation should be decided and start without delay. Also, preparatory release of dam storage, closing endangered traffic roads, stoppage of public transportation etc. should be made without delay. When hazards do arrive, preventive infrastructure should be in full operation such as dams, levees drainage pumps, etc. should be put to a full operation. Assistance to delayed evacuees should be provided depending on local situation.

When disasters do start occurring, emergency response should focus on minimizing losses and preventing expansion of impacts. The first response starts from an individual level and a community level, then the public level such as for rescue of lives, evacuation from damaged areas, etc. Since individuals and communities are at the right spot of disaster occurrence, their immediate actions are decisive to stop progression of disasters. With proper individual actions at the beginning of a disaster, fire can be extinguished, lives are saved, machines are put to safe place, and emergency calls reach right persons right time to stop a concatenation of disasters. Public help is of course important but takes time to arrive at the necessary spot.

Emergency response includes provision of necessary goods and information to evacuation camps such as water, foods, energy, health care, etc. It extends to emergency recovery of lifelines such as water, electricity, communication and transportation, fire extinction, levee breach closings, etc. that should start right after damages were found or reported and the peak of hazards had gone.

The success of such emergency response totally depends on preparedness.

3. Recovery: Recovery includes reconstruction of damaged properties and rehabilitation of livelihood. Once emergency operation nearly comes to an end, full-scale recovery starts in all aspects of lifelines, infrastructure, houses, community life, business and industrial activities. In this recovery stage, risk reduction against next hazards should be included. It should not be just recovery to the original state but improve so as not to repeat the same disaster again. In other words, the principle of recovery should be building back better.

As recovery should start immediately after or even concurrently with the emergency operation if it takes time, the recovery plan needs to be preplanned before a disaster happens. Otherwise, building back better is impossible to be realized as recovery should take place quickly, and once reconstruction starts and people come back, it is extremely hard

Table 2.1 Example of Measures in Each Disaster Management Phase (Slightly modified from ADRC: Website)

Disaster Phase	Earthquake	Flood	Storm (cyclone, typhoon, hurricane)	Landslide
Prevention/ Mitigation	- Seismic design - Retrofitting of vulnerable buildings - Installation of seismic isolation/ seismic response control systems	- Construction of dike - Building of dam - Forestation - Construction of flood control basins/ reservoirs	- Construction of tide wall - Establishment of forests to protect against storms	- Construction of erosion control dams - Construction of retaining walls
Preparedness	- Construction and operation of earthquake observation systems	- Construction and operation of meteorological observation systems	- Construction of shelter - Construction and operation of meteorological observation systems	- Construction and operation of meteorological observation systems
	- Preparation of hazard maps - Food & material stockpiling - Emergency drills - Construction of early warning systems - Preparation of emergency kits			
Emergency Response	- Rescue efforts - First aid treatment - Fire fighting - Monitoring of secondary disaster - Construction of temporary housing - Establishment of tent villages			
Recovery	- Disaster resistant reconstruction - Appropriate land use planning - Livelihood support - Industrial rehabilitation planning			

and costly to redesign settlements and reduce underlying risk afterwards. In case that there are no preplans, yet build back better is aimed, the recovery action will delay till a basic master plan is drawn.

4. Mitigation and prevention: The current society has to be constructed and reformed resilient to disasters. This construction level is most important for underpinning the mitigation and prevention function of disaster management. Mitigation would include the construction that flood may

occur but damage little such as by flood-proof houses, land use regulation (e.g., park, parking lots, high buildings, no residential houses), evacuation ways, shelter construction, etc. Prevention is more direct action against hazards including dams, levees, diversions canals, channel improvements, underground rivers and relocation of residents to safe places.

It is important to note that the main part of risk management is in the stage of development rather than emergency response and crisis management. New risk creation can be avoided and underlying risk can be reduced in this development stage. This is the stage that similar to environmental impact assessment (EIA), risk impact assessment (RIA) should be introduced in a regular administrative licensing or permission process of new structural construction and land development. While EIA is compulsory in many countries, RIA is not necessarily so, although some disaster risk prevention items are included in EIA list.

As shown in Figure 2.6, the four phases of disaster management cycle may be classified into two stages: risk management stage, and crisis management and relief operation stage depending on the focus being an emergency situation or not. Namely, recovery, prevention/mitigation and preparedness are risk management although it should consider a necessity in emergency response. Emergency response including part of preparedness and initial step of recovery is crisis management and relief operation for humanitarian actions after a disaster happens. Especially prevention/mitigation is a matter of development and decides whether society approaches more resilient and sustainable or not. It is often the case that the major budget and international funds are allocated to relief operation after a disaster happens and little for a development phase before a disaster happens. It is politically correct as humanitarian relief operation is by all means necessary to help affected people and recover destroyed houses, lifelines and infrastructure. But needless to say, it is important to "Invest today for a safer tomorrow" as UNISDR used to say (UNISDR: website on Proposal for Climate Change Adaptation). In order to make relief operations connected to build back better towards resilient and sustainable society, it is again stressed that recovery plan should be prepared prior to a disaster.

At any rate, risk management and relief operation should be seamlessly connected. The tendency that risk management is considered low in priority as compared with emergent relief operation should be corrected however high priority politicians tend to put for visible humanitarian efforts.

Instead of disaster management cycle in Figure 2.6, if disaster is considered natural and as an act of God, the management cycle would be as Figure 2.7 where little underlying risk reduction takes place in the prevention and mitigation phase. With mere recovery to the original state without building back better, a similar disaster does repeat again forming a vicious cycle.

Figure 2.7 Vicious cycle of disaster management. To perceive a disaster as an act of God leads to little risk reduction after a disaster and repeats similar disasters again and again.

2.4.2 Hydro-illogical cycle

When some devastating disasters such as floods and droughts happen, the news media report the situation in detail often sensationally, emphasizing the need of immediate help and actions to recover and prevent the repetition in the future. Governments and people respond to them by providing as much help as possible. Societal solidarity is at its peak, many volunteers offer help and opinion, leaders propose more investment, better cooperation and preparedness. But time goes by, such an enthusiasm starts slowly declining, replaced by some other topics and eventually goes back to normal largely forgetting the urgency and necessity of continuous support. News media seek a new target of interest and focus to keep attracting the minds of population. This familiar phenomenon is called a "hydro-illogical cycle" as shown in Figure 2.8 which resembles a hydrological cycle in rainfall-runoff relation with rise, peak and recession.

As long as disaster risk management is concerned, such an illogical cycle of societal interest should not be accepted. Even if the general public irrationally follow the cycle, professionals responsible to disaster risk management should not.

One of the original sources of the word "hydro-illogical cycle" seems the book of I.R. Tannehill "Drought: Its causes and effects", Princeton University Press in 1947, who showed a figure similar to Figure 2.9. It describes a cycle of people's reaction to drought: awareness, concern, panic, then after rainfalls, relief, apathy and another cycle starts again with the next drought. Here panic is the peak of the concern and rain is a rescue that relieves the tension and enables recovery. But soon after relief, people forget about the hardship and apathy prevails making people indifferent to drought. It is quite a familiar reaction cycle of society to any disaster. But in risk conscious society,

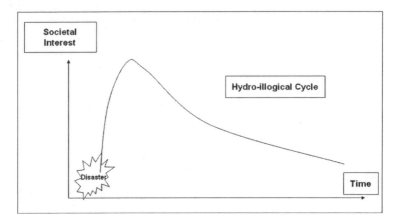

Figure 2.8 Hydro-illogical cycle. Societal interest increases with increase of risk, peaks at a disaster occurrence and slowly declines to none after the severity of a disaster is over. The shape is similar to a hydrological cycle but such an easy-to-heat and easy-to-cool nature is illogical from risk management point of view.

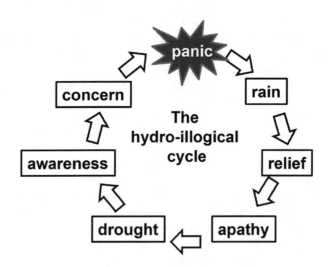

Figure 2.9 Modified from Tannehill's hydro-illogical cycle (Tannehill, 1947), one of the earliest books in which this word was used.

awareness and preparedness should always be kept high to avoid an unanticipated attack of hazards by education, training and building a culture of safety.

A very interesting survey of such phenomena was done by OKADA Norio (Okada, 2002) as in Figure 2.10. He measured the 10-day average area of newspaper articles and the 10-day average water-saving ratio during the

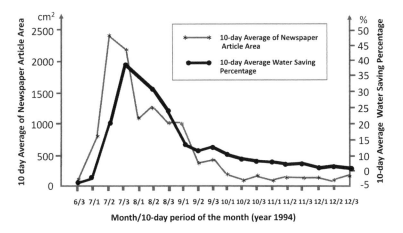

Figure 2.10 Okada's newspaper statistics on drought. Comparison of the 10-day average water saving ratio and the 10-day average area of newspaper articles on drought during the Takamatsu drought from June to December 1994 (Okada, 2002). (Courtesy of OKADA Norio, redrawn by himself for this publication.)

Takamatsu drought from June to December 1994. This clearly indicates the hydro-illogical cycle of the public which disaster managers should not follow.

REFERENCES

ADRC: website www.adrc.asia/publications/TDRM2005/TDRM_Good_Practices/ PDF/PDF-2005e/Chapter2_2.2.pdf (accesses 22 October 2020).

ASU: website on Center for Emergency Management and Homeland Security, Arizona State University. https://cemhs.asu.edu/sheldus (accessed 29 January 2022).

BEH (2011) World Risk Report 2011. Bündnis Entwicklung Hilft. https://collections. unu.edu/eserv/UNU:2046/WorldRiskReport-2011_online_EN.pdf (accessed 2 February 2022).

BEH (2021) World Risk Report 2022. https://reliefweb.int/sites/reliefweb.int/files/ resources/2021-world-risk-report.pdf (accessed 2 February 2022).

Blaikie, Piers, Terry Cannon, Ian Davis and Ben Wisner (1994) *At Risk: Natural Hazards, People Vulnerability and Disasters*, 1st ed. Routledge. http://doi. org/10.4324/9780203428764.

Cutter, Susan L. (2010) Social Science Perspectives of Hazards and Vulnerability Science. In: Tom Beer (ed.), *Geophysical Hazards: Minimizing Risk, Maximizing Awareness*. Springer Science+Business Media, Dordrecht, the Netherlands, Chapter 1, 17–30.

Dartmouth Flood Observatory: website on Global Active Archive of Large Flood Events. http://floodobservatory.colorado.edu/Archives/index.html (accessed 8 October 2019).

EM-DAT: website www.emdat.be/ (accessed 8 October 2019).

IRDR (2014) *Peril Classification and Hazard Glossary (IRDR DATA Publication No. 1)*. Integrated Research on Disaster Risk, Beijing.

A Merriam-Webster (1967) *Webster's Seventh New Collegiate Dictionary*. G. & C. Merriam Co., Springfield, MA.

Munic-Re: website on NatCatSERVICE. www.munichre.com/en/reinsurance/business/non-life/natcatservice/index.html (accessed 8 October 2019).

Okada, Norio (2002) Developing an Indicator of a Community's Disaster Risk Awareness. In: Janos J. Bogardi and Zbigniew W. Kundzewicz (eds.), *Risk, Reliability, Uncertainty, and Robustness of Water Resources*. International Hydrology Series. Cambridge University Press, Cambridge (UNESCO 2004, first published in printed format in 2002), pp. 62–69.

Tannehill, I.R. (1947) *Drought: Its Causes and Effects*. Princeton University Press, Princeton, NJ.

UNDRR: website on DesInventar. www.desinventar.net/ (accessed 29 January 2022).

UNGA (2016) *Report of the Open-ended Intergovernmental Expert Working Group on Indicators and Terminology Relating to Disaster Risk Reduction*. A/71/644, 1 December 2016. www.undrr.org/terminology (accessed 24 January 2022).

UNISDR (2009a) *2009 UNISDR Terminology on Disaster Risk Reduction*. UN International Strategy for Disaster Reduction. www.undrr.org/publication/2009-unisdr-terminology-disaster-risk-reduction (accessed 26 November 2020).

UNISDR (2009b) *Global Assessment Report on Disaster Risk Reduction 2009 (2009 GAR)*. UN International Strategy for Disaster Reduction. www.prevention web.net/english/hyogo/gar/report/documents/GAR_Chapter_1_2009_eng.pdf (accessed 26 November 2020).

UNISDR: website on Proposal for Climate Change Adaptation. www.unisdr. org/2011/docs/donors/donors_proposal_parliamentariansinitiative.pdf (accessed 29 January 2021).

Wisner, Ben, Piers Blaikie, Terry Cannon and Ian Davis (2004) *At Risk: Natural Hazards, People Vulnerability and Disasters*, 2nd ed. Routledge, London and New York.

Chapter 3

Conceptual evolution of UN policies on environment, development and disaster reduction

The main theme of this textbook is the concept of integrated flood risk management (IFRM), which is part of integrated disaster risk management (IDRM) and a subset of integrated water resources management (IWRM). Although both concepts are nothing special and must have been exercised one way or another for a long time in various places in the world, the United Nations' involvement on those with intergovernmental discussion under scientific guidance was rather new and essential for their promotion by international collaboration. The first UN involvement in disaster reduction was only 1990 as the UN International Decade for Natural Disaster Reduction (UNIDNDR), and IWRM was in 1992 when it was adopted as a guiding principle of sustainable freshwater management in Agenda 21, an outcome of the UN Conference on Environment and Development (UNCED). Since then, DRR and IWRM have been made a great stride in the world as the core concepts for managing disaster as well as water resources. Thus this chapter reviews the conceptual evolution of the UN policies on environment, development and disaster reduction which is essential to understanding the concept of IWRM and IFRM.

3.1 FROM STOCKHOLM TO RIO: UN INITIATIVES ON ENVIRONMENT AND DEVELOPMENT

The UN initiatives on environment and development have been evolving since Stockholm Human Environment Conference held in 1972 which was the first UN conference on global issues on human environment. Ever since, a major conference on the topic on environment and development has been organized every 10 years.

DOI: 10.1201/9781003275541-4

3.1.1 Before Stockholm

Before 1972 prior to Stockholm, there were quite a few movements on global environmental issues especially in North America and Europe such as:

- Rachel Carson, a US marine biologist published a book "Silent Spring" (Carson, 1962) and alerted that the progress of the compound contamination by pesticides such as DDT deadly affects environment through food chain all over the world and kill not only pests but also all small insects, without which birds cannot live and spring becomes silent.
- Earth Day was initiated by US Senator Gaylord Nelson who recruited a Stanford University student, Denis Hayes (then in Harvard University) to organize the first Earth Day in the USA on 22 April 1970 when it was said 20 million US people joined. Since 2010 it has been 1 billion people worldwide every year (Earthday: website on History). To honor the Earth was originally proposed by a peace activist John McConnell at UNESCO Conference in 1969 which celebrated the first Earth Day on 21 March (spring equinox) 1970 signed by then UN Secretary-General U Thant. Now it is the Mother Earth Day on 22 April since 2009 (UN: website on International Mother Earth Day).
- In the United States, three ambitious environmental legislation were enacted around 1970, that is, the National Environmental Protection Act (NEPA) of 1969, Clean Air Act Amendments of 1970 and Water Pollution Control Act Amendments (Clean Water Act) of 1972 (Hines, 2013). By the NEPA, the US Environmental Protection Agency (EPA) was established. By the Clean Air Act Amendments of 1970, the automobile (light duty vehicle) emission of carbon monoxide (CO) was ordered to be reduced at least 90% of the level of 1970 by 1975, and same for nitrogen oxides (NOx), at least 90% of the level of 1971 by 1976 (US Senate and House of Representatives, 1970). The 90 percent was really a strong regulation. It was called the Muskie Act as its legislation was proposed and strongly led by Senator Edmund Muskie, who fought the 1968 Presidential Election as Vice President of Hubert Humphrey against Richard Nixon. It was considered the first time in history that the environmental issue became a major political issue.
- Donella H. Meadows, Dennis L. Meadows, Jørgen Randers, and William W. Behrens III published "The Limits to Growth" commissioned by the Club of Rome (1972). They predicted the consequences of a rapidly growing world population and a finite supply of resources by simulating interactions between the Earth's and human systems by the method called "industrial dynamics". It indicated that human growth would be limited by pollution and depletion of resources if human activities continued as it was. The book echoed Thomas Robert Malthus' "An Essay on the Principle of Population" (1798) but indicated what to be done to avert the crisis also (Golub and Townsend, 1977).

3.1.2 Every 10 years UN conferences on environment and development

Responding to such global alerts, the UN organized the Stockholm Conference on Human Environment in 1972 and ever since, every 10 years, the UN has been organizing the follow-up conferences on environment and development. The latest one was in Rio de Janeiro, Brazil, in 2012 and the next will be in 2022.

Although this series started from a global concern on environmental degradation due to developmental activities, it naturally led to conflicts between developed and developing nations as the priority of developing nations was economic development and that of developed was natural environment. This conflict continued to seek a compromise and reached to a notion of sustainable development. The concept of sustainable development is much related with the next series of UN policy evolution, disaster risk reduction. The current conceptual goal of disaster risk reduction is resilience building, which is definitely a necessary condition of sustainable development as without resilience no society can be sustainable. To follow are the chronology of such conceptual evolution of the UN policies on environment and development.

- The UN Conference on Human Environment in Stockholm on 5–16 June 1972.

It was the first UN initiative on global environmental issues. It accepted the Report of the United Nations Conference on Human Environment (A/CONF.48/14/Rev.1). Chapter 1 of the report was "Declaration of the United Nations Conference on Human Environment" which included the proclamation of the following:

1. Man is both creature and moulder of his environment, which gives him physical sustenance and affords him the opportunity for intellectual, moral, social and spiritual growth . . . through the rapid acceleration of science and technology, man has acquired the power to transform his environment in countless ways and on an unprecedented scale . . .
3. . . . We see around us growing evidence of man-made harm in many regions of the earth: dangerous levels of pollution in water, air, earth and living beings; major and undesirable disturbances to the ecological balance of the biosphere; destruction and depletion of irreplaceable resources; and gross deficiencies, harmful to the physical, mental and social health of man, in the man-made environment, particularly in the living and working environment.
4. In the developing countries most of the environmental problems are caused by under-development. Millions continue to have far below the minimum levels required for a decent human existence, deprived of adequate food and clothing, shelter and education, health and

sanitation. Therefore, the developing countries must direct their efforts to development, bearing in mind their priorities and the need to safeguard and improve the environment. For the same purpose, the industrialized countries should make efforts to reduce the gap themselves and the developing countries. In the industrialized countries, environmental problems are generally related to industrialization and technological development.

6. A point has been reached in history when we must shape our actions throughout the world with a more prudent care for their environmental consequences . . . To defend and improve the human environment for present and future generations has become an imperative goal for mankind – a goal to be pursued together with, and in harmony with, the established and fundamental goals of peace and of worldwide economic and social development.

7. . . . The Conference calls upon Governments and peoples to exert common efforts for the preservation and improvement of the human environment, for the benefit of all the people and for their posterity.

Thus it is clear that the concern was not the natural environment but human environment and the problem of developing countries was the main issue. Following such a proclamation, it stated the 26 principles as common convictions. The first two principles express on the basic stands on human right and natural resources of the earth, that is,

Principle 1

Man has the fundamental right to freedom, equality and adequate conditions of life, in an environment of a quality that permits a life of dignity and well-being, and he bears a solemn responsibility to protect and improve the environment for present and future generations. In this respect, policies promoting or perpetuating apartheid, racial segregation, discrimination, colonial and other forms of oppression and foreign domination stand condemned and must be eliminated.

Principle 2

The natural resources of the earth, including the air, water, land, flora and fauna and especially representative samples of natural ecosystems, must be safeguarded for the benefit of present and future generations through careful planning or management, as appropriate.

Then the rest continues as Principles 3–7 on renewable resources, wildlife, non-renewable resources, toxic substances, pollution of the seas; Principles 8–10 on economic and social development, under-development and natural disasters, stability of prices and adequate earning; Principles 11–20 on policies on environment, development, their conflicts, urbanization, demography, and institutional approaches including education and science and technology; Principles 21–26 on international collaboration including UN roles, national responsibility, applicability of standards in each country of different value system, and the need of destruction of nuclear weapons.

In Principle 1, it was clearly stated that the top concern of this declaration was human environment, that is, "an environment that permits a life of dignity and well-being, . . . for the present and the future generations". It was followed by Principles 2–7 which strongly stated the importance of protecting natural environment and resources. But Principles 8–9 again stated the problems of developing countries including disasters as a result of under-development. The main interests of both developed and developing countries seemed carefully treated in a fair balance. Nevertheless, it was true that in follow-up actions in the conference, what prevailed was the concern of natural environment of the globe.

Following those principles, 109 recommendations were accepted as an action plan and one of them was to establish the United Nations Environmental Program (UNEP) for implementation of the declaration and it was executed on the same day. The catchphrase of the Conference was "*Only One Earth*" and the Conference was participated by 113 nations (USSR and East European countries did not as East Germany was not a UN member but West Germany was).

Discussions were from the beginning on the issues of rich and poor, development and environment and the fact that the small fractions of countries consume most of resources and produce vast amount of pollution. There is a precious video where many delegates including the Prime Minister of India Indira Gandhi spoke (UN Audiovisual Library: website on Only One Earth).

• The tenth anniversary of the United Nations Conference on the Human Environment in Nairobi on 10–18 May 1982.

The UNEP was established and the action plan was started implementation but the conflict between environmental protection and development became more apparent in various areas such as deforestation, soil and water degradation, desertification, ozone layer, carbon dioxide, acid rain, disposal of hazardous substances, extinction of animal and plant species, etc. Still more on the issues of human environment in developing countries (UN Documents, 1982). This intensified the conflict between developed and developing countries.

On the occasion of the 10th anniversary of the UNEP, in order to settle such conflict between development and environment, the Director of Environmental Agency in Japan, HARA Bunbei proposed to establish the World Commission on Environment and Development (WCED). It was established in 1983 led by chairwoman Gro Harlem Brundtland of Norway. So the commission was also called as the Brundtland Commission. After a number of meetings, site visits and public hearings in all regions of the world (the last was in Tokyo in February 1987), its final report "*Our Common Future*" was published (WCED, 1987). It proposed the basic concept of "Sustainable development" as the principle to keep harmony between environment and development which became the UN principle to solve the conflict between environment and development. The substantive discussion on sustainable development will be in the next sub-section.

• The United Nations Conference on Environment and Development (UNCED) in Rio de Janeiro on 1992.6.3–14.

It is also known as the Rio Summit or Earth Summit. This conference was meant to put forward the concept of sustainable development proposed by WCED for implementation. After serious preparation meetings in respective subject areas, the Rio Declaration on Environment and Development was agreed that included 27 principles for environment and development to go in harmony, and the action plan "Agenda 21" (UN, 1992) was adopted as a concrete prescription to implement the concept of sustainable development.

Agenda 21 has 40 chapters. Following Chapter 1 Preamble, there are four sections, i.e., Chapters 2–8 Social and economic dimensions, 9–22 Conservation and management of resources for development, 23–32 Strengthening the role of major groups, and 33–40 Means of implementation. The first paragraph of Chapter 1 Preamble states the strong commitment of the UN for cooperative efforts on the integration of environment and development for sustainable development, i.e.,

> 1.1. Humanity stands at a defining moment in history. We are confronted with a perpetuation of disparities between and within nations, a worsening of poverty, hunger, ill health and illiteracy, and the continuing deterioration of the ecosystems on which we depend for our well-being. However, integration of environment and development concerns and greater attention to them will lead to the fulfilment of basic needs, improved living standards for all, better protected and managed ecosystems and a safer, more prosperous future. No nation can achieve this on its own; but together we can – in a global partnership for sustainable development.

The subject on disaster was extensively discussed in Chapter 7 "Promoting sustainable human settlement development". In its Section F "Promoting human settlement planning and management in disaster-prone areas", actions of developing culture of safety, pre-disaster planning and post-disaster reconstruction were stressed. It is impressive to see those important points were already emphasized right after the start of UNIDNDR in 1990 (see Section 3.2.1).

In Chapter 18 "Protection of the quality and supply of freshwater resources: application of integrated approaches to the development, management and use of water resources", the concept of integrated water resource management (IWRM) was presented as the main methodology to implement sustainable freshwater development. In fact, the IWRM was the only methodology adopted in Agenda 21 to realize sustainable development of freshwater resources. But nevertheless, the presented were still rather principles and needed some more time for concrete guidelines to be developed for practice. The IWRM principles were discussed in Dublin in January 1992 prior to Rio Summit as explained in the next chapter, Section 4.1.2.

- The World Summit on Sustainable Development (WSSD) in Johannesburg on 26 August-4 September 2002.

It is also called Earth Summit 2002. The progress of sustainable development was reviewed and WEHAB (water, energy, health, agriculture and biodiversity) was declared priority areas for achieving sustainable development.

- The United Nations Conference on Sustainable Development (or Rio+20) in Rio de Janeiro on 20–22 June 2012.

The final report *"The Future We Want"* (UN, 2012) included the proposals for post-2015 aiming at the post-Hyogo Framework for Action to be adopted at the Third World Conference for Disaster Risk Reduction in Sendai and the Sustainable Development Goals (SDGs) to be adopted by the United Nations Sustainable Development Summit, both planned in 2015. The year 2015 was then an epoch-making year for global sustainability issues when UN assemblies for Sendai Framework for Disaster Risk Reduction (SF), UN SDGs and the 21st Conference of Parties (COP21) in Paris for Framework Convention for Climate Change (FCCC) were held.

- Stockholm+50: a healthy planet for the prosperity of all – our responsibility, our opportunity in Stockholm, Sweden on 2–3 June 2022.

It is announced that based on UN General Assembly's adaptation of a resolution in May 2021, this international environmental meeting will be held to commemorate the 1972 United Nations Conference on the Environment and celebrate 50 years of global environmental action. It aims to "accelerate the implementation of the UN Decade of Action to deliver the Sustainable Development Goals, including the 2030 Agenda, Paris Agreement on climate change, the post-2020 global Biodiversity Framework, and encourage the adoption of green post-COVID-19 recovery plans" (Stockholm+50: website).

Sustainable development proposed by WCED

The World Commission on Environment and Development (WCED) held a number of public hearings in various parts of the world and discussed on how environment and development can go in harmony. The final meeting was held in Tokyo in February 1987 and the final report "Our Common Future" (WCED, 1987) was issued in April 1987, which defined "Sustainable Development" as follows:

> Humanity has the ability to make development sustainable – to ensure that it meets the needs of the present without compromising the ability of future generations to meet their own needs. The concept of sustainability implies limits: not absolute limits, but limitations imposed by the present state of technology and social organization on environmental resources and by the ability of the biosphere to absorb the effect of human activities.

Such concept was in principle not at all new. "Sustainable use" was a common terminology used for a long time in forestry, fishery, groundwater and others which can sustain forever if properly used without excessive use one time. The key concept of this definition is "generational equity", that is, both the present and the future generations should be able to meet their own needs. But in order to satisfy both generations, it is necessary to limit the current use of resources for the future. About this limit, the definition says that it is not a fixed limit but depends on science and technology, societal governance of environment, and the capacity of biosphere.

There were quite many objections for this concept. Some said the word sustainable development is self-contradictory as nothing can develop forever. Some said it was too vague as there were no concrete methodologies presented. The current UN policies seem to give answers to those questions in the following way: Development does not necessarily mean an increase of materialistic products but rather increase of quality of life and environment which has no limit. Once basic material needs are met, however difficult it is, quality improvement is the next goal to be sought, and the development indicator is human well-being instead of materialistic product such as gross domestic product (GDP). The methodology to realize sustainable development is difficult but the sustainable development goals (SDGs) declared by the UN General Assembly in 2015 shows a good list of what to achieve for sustainable development. The concrete methodology to achieve each goal has been under serious trial in each field. There is no exception for integrated flood risk management.

3.1.3 Millennium Development Goals (MDGs) and Sustainable Development Goals (SDGs)

Apart from every 10 years UN conferences on environment and development, in year 2000 celebrating the start of the new millennium, the UN General Assembly set forth the Millennium Development Goals, and in 15 years later, in 2015, Sustainable Development Goals. which quite well accord with recommendations for human environment as well as goals stated in agenda 21. But the most important nature of those goals is that they were set in numerical terms and within time limit.

Millennium Development Goals (MDGs)

At the turn of the Millennium, the UN General Assembly held a Millennium Summit gathered by 189 Member States with 147 heads of State and Government and adopted a resolution, the UN Millennium Declaration on 8 September 2000. This declaration reaffirmed the UN members' "faith in the Organization and its Charter as indispensable foundations of a more peaceful, prosperous and just world" and committed to (1) the fundamental values of freedom, equality, solidarity, tolerance, respect for nature and shared responsibility, (2) peace, security and disarmament, and above all, (3) development and poverty eradication with time-bounded numerical targets as follows (UN, 2000):

- To halve, by the year 2015, the proportion of the world's people whose income is less than one dollar a day and the proportion of people who suffer from hunger and, by the same date, to halve the proportion of people who are unable to reach or to afford safe drinking water (and basic sanitation added in the road map in 2001).
- To ensure that, by the same date, children everywhere, boys and girls alike, will be able to complete a full course of primary schooling and that girls and boys will have equal access to all levels of education.
- By the same date, to have reduced maternal mortality by three quarters, and under-five child mortality by two-thirds, of their current rates.
- To have, by then, halted, and begun to reverse, the spread of HIV/ AIDS, the scourge of malaria and other major diseases that afflict humanity.
- To provide special assistance to children orphaned by HIV/AIDS.
- By 2020, to have achieved a significant improvement in the lives of at least 100 million slum dwellers as proposed in the "Cities Without Slums" initiative.

Here there was no numerical target on disaster reduction. The reasons considered were because disaster losses, either human or economic, could not be accurately measured in many nations so that even if some numerical targets were set, the way to achieve them could not be formulated nor its progress monitored.

Following this, in September 2001, the UN General Assembly adopted the road map of achieving these targets as A/56/326 (UN, 2001) which were called as the Millennium Development Goals (MDGs). For each goal, a logo was assigned as in Figure 3.1.

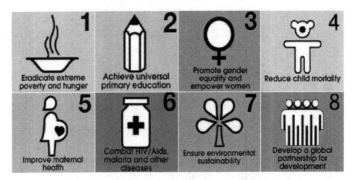

Figure 3.1 The Millennium Development Goals (MDGs), 2000. (UN: website on MDGs)

There were 18 targets and 48 indicators (later rearranged into 21 and 60, respectively for monitoring purpose) which were classified into eight goals as follows:

Goal 1. Eradicate extreme poverty and hunger,
Goal 2. Achieve universal primary education,
Goal 3. Promote gender equality and empower women,
Goal 4. Reduce child mortality,
Goal 5. Improve maternal health,
Goal 6. Combat HIV/AIDS, malaria and other diseases,
Goal 7. Ensure environmental sustainability, and
Goal 8. Develop a global partnership for development.

Under those goals, targets and indicators were rearranged into 21 and 60 for monitoring purpose after some iterations of discussions (UNICEF: website on MDGs monitoring). To set the time-bounded numerical targets and monitor them were epoch-making in UN initiatives. Since all governments seriously committed and made considerable efforts to carry them out jointly with various international organizations, many targets were well achieved by 2015. The final assessment "The Millennium Development Goals Report 2015" (UN, 2015a) of the UN indicated such as:

- In 1990, nearly half of the population in the developing world lived on less than $1.25 a day; that proportion dropped to 14 percent in 2015.
- The developing regions' primary school net enrolment rate has reached 91 percent in 2015, up from 83 percent in 2000.

- In Southern Asia, only 74 girls were enrolled in primary school for every 100 boys in 1990. Today, 103 girls are enrolled for every 100 boys.
- The global under-five mortality rate has declined by more than half, dropping from 90 to 43 deaths per 1000 live births between 1990 and 2015.
- Since 1990, the maternal mortality ratio has declined by 45 percent worldwide, and most of the reduction has occurred since 2000.
- In 2015, 91% of the global population is using an improved drinking water source, up from 76% in 1990 – the target was met 5 years ahead of the 2015 deadline.

Thus, the MDGs achievements were outstanding in many cases. Nevertheless, they were uneven in the world and not complete at all. The efforts were therefore necessary to be continued.

Sustainable Development Goals (SDGs)

Responding to such needs, the UN agencies, international professional organizations and other stakeholders got together and discussed in length the succeeding program. The final meeting was the UN General Assembly which adopted a resolution "Transforming our world: the 2030 Agenda for Sustainable Development" (A/70/L.1) on 25 September 2015 (UN, 2015b). Logos assigned were as in Figure 3.2.

Figure 3.2 Sustainable Development Goals (SDGs), 2015. (UN: website on SDGs)

The successor of MDGs is Sustainable Development Goals (SDGs). The SDGs have 17 goals and 169 targets, namely:

Goal 1. End poverty in all its forms everywhere.

Goal 2. End hunger, achieve food security and improved nutrition and promote sustainable agriculture.

Goal 3. Ensure healthy lives and promote well-being for all at all ages.

Goal 4. Ensure inclusive and equitable quality education and promote lifelong learning opportunities for all.

Goal 5. Achieve gender equality and empower all women and girls.

Goal 6. Ensure availability and sustainable management of water and sanitation for all.

Goal 7. Ensure access to affordable, reliable, sustainable and modern energy for all.

Goal 8. Promote sustained, inclusive and sustainable economic growth, full and productive employment and decent work for all.

Goal 9. Build resilient infrastructure, promote inclusive and sustainable industrialization and foster innovation.

Goal 10. Reduce inequality within and among countries.

Goal 11. Make cities and human settlements inclusive, safe, resilient and sustainable.

Goal 12. Ensure sustainable consumption and production patterns.

Goal 13. Take urgent action to combat climate change and its impacts.

Goal 14. Conserve and sustainably use the oceans, seas and marine resources for sustainable development.

Goal 15. Protect, restore and promote sustainable use of terrestrial eco-systems, sustainably manage forests, combat desertification, and halt and reverse land degradation and halt biodiversity loss.

Goal 16. Promote peaceful and inclusive societies for sustainable development, provide access to justice for all and build effective, accountable and inclusive institutions at all levels.

Goal 17. Strengthen the means of implementation and revitalize the Global Partnership for Sustainable Development.

The number of goals increased from MDGs to SDGs and in essence, the MDGs are all included in the new SDGs. The 1st and 2nd goals of SDGs "end poverty" and "end hunger" are the 1st goal of MDGs "eradicate poverty and hunger". Similarly, the 3rd of SDGs "good health and well-being" includes the 4th, 5th and 6th goals of MDGs "child mortality", "maternal health" and "HIV and malaria", then the 2nd, 3rd and 7th of MDGs "primary education", "gender equality" and "environmental sustainability" are followed by 4th, 5th and 15th of SDGs and part of 7th of MDGs "water and sanitation" is 6th of SDG. Also, the important goals "partnership and collaboration" to achieve all other goals are set in common at the last MDG 8th and the last SDG 17th.

Looking at the disaster risk reduction theme, it is considered as a cross-cutting theme necessary to achieve many goals such as the 1st end poverty, 2nd end hunger and 13th combat climate change and especially in the 11th goal, "make cities and human settlements inclusive, safe, resilient and sustainable". In the 11th goal, there are two targets 11.5 and 11.6 set, namely,

Target 11.5 By 2030, significantly reduce the number of deaths and the number of people affected and substantially decrease the direct economic losses relative to global GDP caused by disasters, including water-related disasters, with a focus on protecting the poor and people in vulnerable situations.

Target 11.6 By 2020, substantially increase the number of cities and human settlements adopting and implementing integrated policies and plans towards inclusion, resource efficiency, mitigation and adaptation to climate change, resilience to disasters, and develop and implement, in line with the Sendai Framework for Disaster Risk Reduction 2015–2030, holistic disaster risk management at all levels.

Those targets are not numerical but strong words "significantly" and "substantially" are used following the Sendai Framework for DRR agreed in Sendai six months before in March 2015. The basic reason of no numerical targets is its difficulty in measuring human as well as economic losses and formulating effective risk reduction plans.

It is important to note, however, that such global targets can only be achieved starting from tractable practical actions in local communities such as preparing historical inundation maps. If such a small real thing is not possible, no large targets can ever be reached and the goals stated are just a big talk.

3.2 FROM YOKOHAMA TO SENDAI: UN INITIATIVES ON DISASTER REDUCTION

Apart from UN initiatives on environment and management, UN initiatives on disaster reduction have been taking place as another line of major UN policies. Environment and development have been a conflicting issue and as seen in the previous section, the main compromise should be in between developed and developing countries. It is absolutely necessary to eradicate poverty to stop environmental degradation and protect people from degraded environment. The disaster reduction from natural hazards has been, on the other hand, developed free from conflict between developed and developing countries although poverty is still a major issue for disaster reduction as it increases vulnerability to form a vicious cycle of a disaster risk-poverty nexus. Two initiatives seem to merge under the concept of sustainable development because building societal resilience to disasters is a necessary condition for sustainable development.

3.2.1 United Nations International Decade for Natural Disaster Reduction (UNIDNDR) (1990–1999)

Dr. Frank Press, the President of the U.S. National Academy of Sciences launched the idea of the international Decade in 1984 and on 11 December 1987 the United Nations General Assembly adopted Resolution 42/169 to designate the 1990s as an International Decade for Natural Disaster Reduction (IDNDR) in which the world community joined to cooperate in natural disaster reduction. The basic idea behind this proclamation of the Decade was the unacceptable and rising levels of losses that disasters continue to incur on the one hand, and the existence, on the other hand, of a wealth of scientific and engineering know-how which could be effectively used to reduce losses resulting from disasters.

Yokohama strategy

The UN World Conference on Natural Disaster Reduction, later referred as the 1st World Conference on Disaster Reduction, which was part of a mid-term review of Decade activities, was held in Yokohama (Japan), 23–27 May 1994 and adopted "The Yokohama Strategy for a Safer World: Guidelines for Natural Disaster -Prevention, Preparedness and Mitigation and its Plan of Action-" (UNISDR, 1994). The strategy was made up of ten principles listed in the following, which became the bases of later actions such as Hyogo Framework for Action adopted in 2005. (Underlines are added by the author.)

1. Risk assessment is a required step for the adoption of adequate and successful disaster reduction policies and measures.
2. Disaster prevention and preparedness are of primary importance in reducing the need for disaster relief.
3. Disaster prevention and preparedness should be considered integral aspects of development policy and planning at national, regional, bilateral, multilateral and international levels.
4. The development and strengthening of capacities to prevent, reduce and mitigate disasters is a top priority area to be addressed during the Decade so as to provide a strong basis for follow-up activities to the Decade.
5. Early warnings of impending disasters and their effective dissemination using telecommunications, including broadcast services, are key factors to successful disaster prevention and preparedness.
6. Preventive measures are most effective when they involve participation at all levels, from the local community through the national government to the regional and international level.

7. Vulnerability can be reduced by the application of proper design and patterns of development focused on target groups, by appropriate education and training of the whole community.
8. The international community accepts the need to share the necessary technology to prevent, reduce and mitigate disaster; this should be made freely available and in a timely manner as an integral part of technical cooperation.
9. Environmental protection as a component of sustainable development consistent with poverty alleviation is imperative in the prevention and mitigation of natural disasters.
10. Each country bears the primary responsibility for protecting its people, infrastructure, and other national assets from the impact of natural disasters. The international community should demonstrate strong political determination required to mobilize adequate and make efficient use of existing resources, including financial, scientific and technological means, in the field of natural disaster reduction, bearing in mind the needs of the developing countries, particularly the least developed countries.

3.2.2 United Nations Office for the Coordination of Humanitarian Affairs (UNOCHA)

In 1991, the UN General Assembly passed Resolution 46/182 (UN, 1991) that aimed to strengthen the international community's collective effort, particularly the United Nations System, to make emergency response more effective. Based on this resolution, OCHA was established as an arm of the UN Secretariat that is responsible for bringing together humanitarian actors to ensure coherent response to emergencies with a framework within which each actor can contribute to the overall response effort.

There are six key humanitarian actors in the UN humanitarian system. They are UNHCR (UN High Commissioner for Refugees) established in 1950, WFP (World Food Programme) established in 1961, UNICEF (UN Children's Fund) established in 1946 which was until 1953 called UN International Children's Emergency Fund, FAO (Food and Agriculture Organisation) established in 1945, WHO (World Health Organization) established in 1945 and UNDP (UN Development Program) established in 1965. Each of them was established by a separate treaty, with its own governance mechanism so that effective coordination was necessary for them to work together in emergency. It was 1971 when the first attempt was made to provide an efficient coordination by establishing the Office of the UN Disaster Relief Coordinator (UNDRO). But its coordination power was too weak to meet the requests from member countries, especially revealed at the time of the Gulf War (2 August 1990–28 February 1991) and in December 1991, the UNOCHA was established with more streamlined mandates for coordination (Reindorp and Wiles, 2001).

Resolution 46/182 is still in force today and serves as the basis of the OCHA mandate. It outlines four fundamental principles governing humanitarian relief: humanity, neutrality, impartiality and independence.

OCHA's mission is to mobilize and coordinate effective and principled humanitarian action in partnership with national and international actors in order to alleviate human suffering in disasters and emergencies; advocate for the rights of people in need; promote preparedness and prevention; and facilitate sustainable solutions. They have been working for refugees in many conflict areas such as Somalia, Afghanistan, Syria, etc. and disaster areas such as the Indian Ocean tsunami that occurred in 2004, Haiti earthquake in 2010, Pakistan floods in 2010 and many other disaster sites (Ferris, 2010; UNOCHA, 2011).

3.2.3 United Nations Disaster Risk Reduction (UNDRR)

The UN Office for Disaster Risk Reduction (UNDRR) is the one renamed from the original UN Office for International Strategy for Disaster Reduction (UNISDR) in May 2019. The UNISDR was established in year 2000 as a UN inter-agency task force and inter-agency-secretariat for disaster reduction, under the direct authority of the Under-Secretary-General for Humanitarian Affairs. When the UN International Decade for Natural Disaster Reduction (UNIDNDR) was going to end in 1999, it was proposed by the UN Secretary-General endorsed by the GA (A/RES/54/219) taking note of Economic and Social Council resolution 1999/63 of 30 July 1999. This GA also decided to maintain the observance of the International Day for Disaster Reduction on the second Wednesday of October (UNDRR: website on about UNDRR).

The UN International Strategy for Disaster Reduction (UNISDR) was a strategic framework, aiming to guide and coordinate the efforts of a wide range of partners to achieve substantive reduction in disaster losses and build resilient nations and communities as an essential condition for sustainable development.

In March 2018, Ms. MIZUTORI Mami, a Japanese diplomat, was appointed to United Nations Assistant Secretary-General and Special Representative of the Secretary-General for Disaster Risk Reduction in the United Nations Office for Disaster Risk Reduction. Within the UN system, UNISDR changed its name to United Nations Disaster Risk Reduction (UNDRR) on 1 May 2019 (UNDRR: website on Home).

The UNISDR led by Director Salvano Briceno made a great effort to advocate the importance of exposure and vulnerability reduction of society. Their symbolic slogans were quite persuasive such as cited from then UNISDR home page (UNISDR: website on What Is Disaster Risk Reduction?):

There is no such thing as a "natural" disaster, only natural hazards.

Disasters often follow natural hazards. A disaster's severity depends on how much impact a hazard has on society and the environment. The scale of the impact in turn depends on the choices we make for our lives and for our environment. These choices relate to how we grow our food, where and how we build our homes, what kind of government we have, how our financial system works and even what we teach in schools. Each decision and action makes us more vulnerable to disasters – or more resilient to them.

Disaster risk reduction is about choices.

Disaster risk reduction is the concept and practice of reducing disaster risks through systematic efforts to analyze and reduce the causal factors of disasters. Reducing exposure to hazards, lessening vulnerability of people and property, wise management of land and the environment, and improving preparedness and early warning for adverse events are all examples of disaster risk reduction.

Disaster risk reduction is everyone's business.

Disaster risk reduction includes disciplines like disaster management, disaster mitigation and disaster preparedness, but DRR is also part of sustainable development. In order for development activities to be sustainable they must also reduce disaster risk. On the other hand, unsound development policies will increase disaster risk – and disaster losses. Thus, DRR involves every part of society, every part of government, and every part of the professional and private sector.

Those concepts were in accordance with those of "At Risk" (Wisner et al., 2004). There is a good discussion by Jonatan Lassa (CDU: website on Is There Such a Thing as "Natural" Disasters?). First notion of societal vulnerability in modern history seems to originate in discussions after the Great Lisbon Earthquake in 1755 as mentioned at the beginning of Chapter 1.

3.2.4 Hyogo Framework for Action (HFA) 2005–2015

UNISDR organized the World Conference on Disaster Reduction, held in Kobe, Hyogo Prefecture, Japan on 18–22 January 2005 and adopted the Hyogo Framework for Action (HFA) (UNISDR, 2005). Later this conference became called the 2nd World Conference on Disaster Risk Reduction following the 1st one in Yokohama in 1994.

The HFA was formulated as a ten-year plan of action to protect lives and livelihoods against disasters in response to international concern about the growing impacts of disasters on individuals, communities and national development. Based on careful study of trends in disaster risks and practical experience in disaster risk reduction since the Yokohama Strategies, and subjected to intensive prior negotiations, the HFA was finally adopted by 168 governments in Hyogo in 2005. UNISDR served as the focal point for the implementation of the HFA.

The HFA: "Hyogo Framework for Action 2005–2015: Building the resilience of nations and communities to disasters" put forward the following priorities (UNISDR, 2005):

Priorities for action

Drawing on the conclusions of the review of the Yokohama Strategy, and on the basis of deliberations at the World Conference on Disaster Reduction and especially the agreed expected outcome and strategic goals, the Conference has adopted the following five priorities for action:

1) Ensure that disaster risk reduction is a national and a local priority with a strong institutional basis for implementation.
2) Identify, assess and monitor disaster risks and enhance early warning.
3) Use knowledge, innovation and education to build a culture of safety and resilience at all levels.
4) Reduce the underlying risk factors.
5) Strengthen disaster preparedness for effective response at all levels.

It is important to note that those priorities were adopted by 168 governments by themselves. In this sense, priority 1 is especially important, namely, they agreed that disaster risk reduction is a national and a local priority. Each government, national or local, should consider it as its own responsibility and establish a strong institutional basis for implementation. As quite often such self-consciousness was weak, this policy accelerated governments to be aware of its importance and form proper institutional arrangements.

The other notable characteristic of this framework was that all priorities were for risk reduction before a disaster happens. Institutional arrangements, risk assessment and early warning, science and education for safety culture and resilience, reduce underlying risk, and preparedness are all what to be done before a disaster happens. Here a strong commitment of ISDR to reduce vulnerability was clearly seen and quite well stated. The most important and very difficult priority was number 4, to reduce underlying risk. This meant that disaster risk is created at the stage of development by changing social, industrial and environmental conditions especially land use. For development, however, as cost-effectiveness is the most important constraint, it is difficult to pay enough attention to risk reduction. As a result, it becomes often the case that after creating risk, society pays much money to reduce risk or recover from a disaster. According to Radu (2012), Gabriel Ionita, an irrigation specialist of the World Bank said that 1 dollar prevention saves 7 dollars after a catastrophic disaster.

3.2.5 Sendai Framework for disaster risk reduction (SF) 2015–2030

Ten years later after Hyogo, the United Nations 3rd World Conference on Disaster Risk Reduction was held in Sendai on 16–18 March 2015 and, based on extensive review of the progress in the HFA, agreed on (A/CONF.224/CRP.1) the new action plan Sendai Framework for Disaster Risk Reduction (Sendai Framework or simply SF) with the following new priorities (UNISDR, 2015):

1) Understanding disaster risk.
2) Strengthening disaster risk governance to manage disaster risk.
3) Investing in disaster risk reduction for resilience.
4) Enhancing disaster preparedness for effective response and to "Build Back Better" in recovery, rehabilitation and reconstruction.

Those priorities of the SF are not a replacement of the HFA but to support the HFA for further strengthening and promotion. The review of the HFA revealed that none was enough achieved to be ready to be replaced by any others. But, additional efforts were found necessary such as understanding risk, governance, investment, and the actions after a disaster. Improving those aspects, the HFA was considered to progress better. During the review of the HFA, it was found that Priority 1 to ensure that disaster risk reduction was a national and a local priority with strong institutional support was relatively well implemented and Priority 4 to reduce underlying risk was least achieved. Such results were quite reasonable because institutional arrangements such as establishing a national platform were rather easy to implement as little budget was necessary, but reduction of underlying risk during the development stage needed much budget and difficult due to the short-term cost recovery pressure.

Reflecting such review results, the SF emphasizes science and technology, governance, investment and building back better to augment and strengthen the Hyogo priorities. The emphasis on science and technology supports understanding risk which is the basis of all risk reduction activities. Building back better is new in Sendai and unique as it addresses the matter after disaster rather than before disaster happens. But it makes sense because however seriously risk reduction is implemented, disasters do occur and recovery and reconstruction take place. It is therefore a great opportunity for society to reduce underlying risk and improve settlements not to repeat the same tragedy again in the future.

The expected outcome, the goal and seven global targets

In the SF the expected outcome, the goal and seven global targets are stated as follows. First, about the expected outcome:

> Building on the Hyogo Framework for Action, the present Framework aims to achieve the following outcome over the next 15 years:
>
> The substantial reduction of disaster risk and losses in lives, livelihoods and health and in the economic, physical, social, cultural and environmental assets of persons, businesses, communities and countries.

Here the word "substantial reduction" is used as a result of long discussion aiming at a numerical target. "Substantial" is a qualitative word but it is a strong word. Although a numerical target was not realized, this statement together with the seven global targets (a)–(g) listed below shows a considerable progress.

Next, the goal is stated as:

> To attain the expected outcome, the following goal must be pursued:
>
> Prevent new and reduce existing disaster risk through the implementation of integrated and inclusive economic, structural, legal, social, health, cultural, educational, environmental, technological, political and institutional measures that prevent and reduce hazard exposure and vulnerability to disaster, increase preparedness for response and recovery, and thus strengthen resilience.

This goal statement is important. It starts from "prevent new disaster risk" and then "reduce existing disaster risk". Although it is a general statement, its meaning is significant. As was found in the review of the HFA, Priority 4 "Reduce the underlying risk factors" was most difficult to implement because it needs to be done in the development stage. This goal now aims for the reduction of disaster risk at the beginning of the planning stage of development. It is really a vicious cycle to develop new risk and try to reduce the created risk afterwards. Creation of underlying risk should be avoided in the planning stage which is much cheaper than to create new risk and then reduce the created risk. This is why as stated in Section 2.4.1, a pre-assessment system "risk impact assessment (RIA)" is necessary to be institutionalized and applied whenever developmental plan is proposed.

Finally, about the global targets:

> To support the assessment of global progress in achieving the outcome and goal of the present Framework, seven global targets have been agreed.
>
> (a) Substantially reduce global disaster mortality by 2030, aiming to lower the average per 100,000 global mortality rate in the decade 2020–2030 compared to the period 2005–2015;

(b) Substantially reduce the number of affected people globally by 2030, aiming to lower the average global figure per 100,000 in the decade 2020–2030 compared to the period 2005–2015;

(c) Reduce direct disaster economic loss in relation to global gross domestic product (GDP) by 2030;

(d) Substantially reduce disaster damage to critical infrastructure and disruption of basic services, among them health and educational facilities, including through developing their resilience by 2030;

(e) Substantially increase the number of countries with national and local disaster risk reduction strategies by 2020;

(f) Substantially enhance international cooperation to developing countries through adequate and sustainable support to complement their national actions for implementation of the present Framework by 2030;

(g) Substantially increase the availability of and access to multi-hazard early warning systems and disaster risk information and assessments to people by 2030.

What is the significance of these seven global targets? Does this strange numerical expression make sense: "aiming to lower the average per 100,000 global mortality rate in the decade 2020–2030 compared to the period 2005–2015"? It is indeed sensible:

1) Because the world population is increasing, the absolute number of disaster casualties may increase. But this target requests that at least "per 100,000" rate should be made "substantially" decreased.

2) It also specifies the dates to be achieved by the year 2030, measured on the global average "in the decade 2020–2030 compared to the period 2005–2015". These specifications make what to be measured clear without ambiguity.

3) The "global" basis is especially important because such measurements should be made everywhere in the world. Global includes all nations in the world without exception. It would be difficult for quite many countries to establish an administrative scheme to be able to count disaster damages, human and economic losses, in good accuracy but otherwise the requirement of "global" cannot be achieved.

4) The 2005–2015 basis refers the past. Therefore the latest archive should be available. If such data are not available in any single nation or region, strictly speaking, the "global" target cannot be achieved. It is an aggregation of strong commitment made by each nation.

5) In fact, however, the target is "aimed to lower" and not just "to lower" which may be a compromise to admit that in some parts of the world

some measurements may not be available and some extraordinary events may happen to exceed the target.

Such targets cover (a) mortality, (b) affected people, and (c) economic losses in GDP. Other targets (d) critical infrastructure and basic services (e) national and local strategies, (f) ODA and (g) early warning have no numerical specifications but the direction is clear to achieve "substantial" improvements.

REFERENCES

Carson, Rachel (1962) *Silent Spring*. Houghton Mifflin Company, Boston, MA.

Earthday: website on History. www.earthday.org/history/ (accessed 22 September 2021).

Ferris (2010) *Earthquakes and Floods: Comparing Haiti and Pakistan*. www.brookings.edu/wp-content/uploads/2016/06/0826_earthquakes_floods_ferris.pdf (accessed 14 November 2020).

Golub, Robert and Joe Townsend (1977) Malthus, Multinationals and the Club of Rome. *Social Studies of Science*, 7(2), 201–222. www.jstor.org/stable/284875 (accessed 25 January 2021).

Hines, N. William (2013) History of the 1972 Clean Water Act: The Story Behind How the 1972 Act Became the Capstone on a Decade of Extraordinary Environmental Reform. *Journal of Energy and Environmental Law, The George Washington University*, 4(2), Summer. https://gwjeel.com/wp-content/uploads/2013/10/4-2-hines.pdf (accessed 28 December 2021).

Lassa, Jonatan at CDU: website on Is There Such a Thing as "Natural" Disasters? www.cdu.edu.au/launchpad/research-impact/there-such-thing-natural-disasters (accessed 28 August 2020).

Malthus, Thomas Robert (1798) *An Essay on the Principle of Population*.

Meadows, Donella H., Dennis L. Meadows, Jørgen Randers and William W. Behrens III (1972) *The Limits to Growth*. Commissioned by the Club of Rome. Universe Books, New York.

Radu, Oana (2012) *The World Bank: Investing 1 Dollar in Prevention Saves 7 Dollars Spent After a Catastrophic Event*. www.xprimm.com/The-World-Bank-Investing-1-dollar-in-prevention-saves-7-dollars-spent-after-a-catastrophic-event-articol-2,10,25–2573.htm (accessed 15 November 2020).

Reindorp and Wiles (2001) *Humanitarian Coordination: Lessons from Recent Field Experience*. A Study Commissioned by the Office for the Coordination of Humanitarian Affairs (OCHA). www.unhcr.org/3bb04e232.pdf (accessed 28 August 2020).

Stockholm+50: website www.stockholm50.global/ (accessed 5 January 2021).

UN (1972) *Report of the United Nations Conference on Human Environment* (A/CONF.48/14/Rev.1). https://undocs.org/en/A/CONF.48/14/Rev.1 (accessed 10 November 2020).

UN (1991). https://undocs.org/A/RES/46/182 (accessed 14 November 2020).

UN (1992) *Agenda 21. United Nations Conference on Environment & Development*. Rio de Janeiro, Brazil, 3–14 June 1992. https://sustainabledevelopment. un.org/outcomedocuments/agenda21 (accessed 26 November 2020).

UN (2000) *United Nations Millennium Declaration*. UN 55th General Assembly on 18 September 2000, A/RES/55/2. http://undocs.org/A/RES/55/2 (accessed 4 September 2020).

UN (2001) *Road Map Towards the Implementation of the United Nations Millennium Declaration*. Follow-up to the Outcome of the Millennium Summit, A/56/326. https://undocs.org/A/56/326 (accessed 4 September 2020).

UN (2012) *The Future We Want. Outcome Document of the United Nations Conference on Sustainable Development*. Rio de Janeiro, Brazil, 20–22 June 2012.

UN (2015a) *The Millennium Development Goals Report 2015*. https://www.unilibrary.org/content/books/9789210574662/read (accessed 4 September 2020).

UN (2015b) *Transforming Our World: The 2030 Agenda for Sustainable Development*. (A/70/L.1), 25 September 2015. www.un.org/ga/search/view_doc. asp?symbol=A/RES/70/1&Lang=E (accessed 4 September 2020).

UN: website on International Mother Earth Day. www.un.org/en/observances/earthday/background (accessed 22 September 2021).

UN: website on MDGs. www.un.org/millenniumgoals/mdgmomentum.shtml (accessed 3 September 2020).

UN: website on SDGs. https://sdgs.un.org/goals (accessed 3 September 2020).

UN Audiovisual Library: website on Only One Earth. www.unmultimedia.org/avlibrary/asset/2408/2408654/ (accessed 10 November 2020).

UN Documents (1982) *Nairobi Declaration*. www.un-documents.net/nair-dec.htm (accessed 24 September 2021).

UNDRR: website on About UNDRR. www.undrr.org/about-undrr/history (accessed 15 November 2020).

UNDRR: website on Home. www.undrr.org/ (accessed 28 August 2020).

UNICEF: website on MDGs Monitoring. www.unicef.org/statistics/index_24304. html#:~:text=The%20Millennium%20Development%20Goals%20 (MDGs,are%20expected%20to%20be%20met (accessed 9 October 2020).

UNISDR (1994) *The Yokohama Strategy for a Safer World: Guidelines for Natural Disaster—Prevention*. Preparedness and Mitigation and Its Plan of Action. http://dpanther.fiu.edu/sobek/FI13042666/00001 (accessed 4 September 2020).

UNISDR (2005, January) *The Hyogo Framework for Action (HFA)*. 2nd World Conference on Disaster Risk Reduction, Kobe, Hyogo. A/CONF.206/6. www.unisdr. org/eng/hfa/hfa.htm (accessed 28 August 2020).

UNISDR (2015) *Sendai Framework for Disaster Risk Reduction 2015–2030*. www. preventionweb.net/files/43291_sendaiframeworkfordrren.pdf (accessed 28 August 2020).

UNISDR: website on What Is Disaster Risk Reduction? https://eird.org/americas/we/ what-is-disaster-risk-reduction.html (accessed 24 January 2022).

UNOCHA (2011) *OCHA Annual Report 2010*. https://reliefweb.int/sites/reliefweb. int/files/resources/Full_Report_1118.pdf (accessed 14 November 2020).

US Senate and House of Representatives (1970) *PUBLIC LAW 91–604*, 31 December 1970. www.govinfo.gov/content/pkg/STATUTE-84/pdf/STATUTE-84-Pg1676. pdf (accessed 28 December 2021).

WCED (1987) *Our Common Future. Report of the World Commission on Environment and Development.* Oxford University Press, Oxford. https://sustainabledevelopment.un.org/content/documents/5987our-common-future.pdf   (accessed 10 November 2020).

Wisner, Ben, Piers Blaikie, Terry Cannon and Ian Davis (2004) *At Risk: Natural Hazards, People Vulnerability and Disasters.* Routledge, London and New York, 2nd ed.

Chapter 4

An integrated approach to water resources and flood risk

In this chapter, integrated flood risk management (IFRM) will be presented as part of integrated disaster risk management (IDRM) and integrated water resources management (IWRM). Integration is necessary because flood management is not an isolated activity in society but connected with the total societal system especially development of human settlement and industrial activities such as infrastructure development, institutional, health, environmental and other economic activities. This is why flood risk management should be considered as part of IWRM and part of societal development. This chapter first introduces the basic concept of IWRM and then the concept of IFRM and finally its implementation procedure. The concrete cases will be introduced through Japanese experiences in Chapter 5.

4.1 INTEGRATED WATER RESOURCES MANAGEMENT (IWRM)

4.1.1 Introduction

As was mentioned in Section 3.1.2, IWRM was the methodology, in fact the only methodology, presented in Agenda 21 to achieve sustainable development of freshwater resources. This section looks into its conceptual framework that relates IWRM with IFRM and other related concepts, and then a concrete methodology applicable to practice will be presented.

IWRM, IFRM and related concepts in a conceptual framework

IWRM and IDRM are the parents of IFRM, the main theme of this textbook, but there are several related terminologies and concepts around them. This section discusses their relationship in conceptual framework using Figure 4.1. Here the discussion on with or without "integrated" is omitted and treated all with "integrated" for simplicity. Some of the integration issues will be discussed in Sections 4.1.3 and 4.2.2.

DOI: 10.1201/9781003275541-5

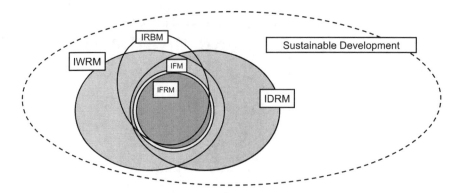

Figure 4.1 Schematic relation among IWRM, IDRM, IRBM, IFM, IFRM and sustainable development. Integrated Flood Risk management (IFRM) is subset of Integrated Disaster Risk Management (IDRM) and Integrated Water Resources Management (IWRM). All those are for sustainable development.

The largest oval in Figure 4.1 is the domain of sustainable development (SD) which contains all living things, including human society, and related geophysical environment. The oval itself may be considered floating in the universe. Whether this domain moves to the state of global sustainability or not depends on the activities within it. All societal actions including those water and disaster management efforts IFM, IFRM, IRBM, IDRM and IWRM are, positive or negative, developmental actions and should accord with principles of sustainable development if society seeks for global sustainability. Their quality determines the quality of development, whether it moves society to resilient and sustainable or remains vulnerable and chaotic.

There are two large ovals within SD, representing the water management sector by IWRM and the disaster management sector by IDRM. They have some mutually exclusive specialties with each other but share a large part in common. In that common part, IFM and IFRM circles are situated where IFRM is drawn within IFM as FM includes FRM. Both IFRM and IFM are totally included in IDRM as neither can be free from disaster risk.

FRM focuses on flood risk, the negative aspect of flood, stressing the importance of actions before it turns into a flood disaster. On the other hand, flood management (FM) focuses on both positive and negative aspects of flood. Therefore, FRM is a subset of FM, and IFRM is a subset of IFM as well. However, FRM without considering its positive aspect is not IFRM, and FM without considering negative aspect is not IFM, either. Therefore, in an integrated approach, barriers between them need to disappear and all sectors attain better coordinated, higher and balanced achievements. Under such process, the distinction between IFRM and IFM becomes practically meaningless.

In fact, IFRM and IFM are often used as nearly a synonym. In the Associated Program on Flood Management (APFM), a joint program of the World Meteorological Organization (WMO) and Global Water Partnership (GWP) (WMO, 2009) as discussed in Section 4.2.4. used the word IFM while it is nearly fully focusing on flood risk.

IRBM is a concept of geographical classification which may include activities of more than water and disaster such as regional developmental issues in general as seen in the case of the Tennessee Valley Authority (TVA) in the US explained in the next section. So here it is shown as an oval having an extra part independent of disaster and IWRM. But as long as river basin is a focus and a bound of consideration, water is a key and no action is free from water consideration. In fact, there may be a different view on this classification which would be interesting to discuss with some practical issues.

IFM and IFRM have the area outside of IRBM because flood management is impossible in some cases if its areal consideration is restricted to each river basin exclusively. River basin is obviously a basic unit of hydrology so that it is the basis of flood management as well as any other water resources management including navigation, hydro-power generation, groundwater, water contamination, ecological and environmental matters, etc. But there are many inter-basin issues such as, in addition to wide ranges of information exchange needs, inter-basin sewerage system, underground rivers, evacuation and other disaster management activities, etc., especially in large cities where administrative boundaries do not coincide with river boundaries.

The most difficult river basin management is the case of transboundary rivers divided by two or more nations. According to Giordano and Wolf (2003), there are 263 international basins (including rivers and lakes) sheared by 145 countries where about 40% of the world's population live, covering nearly half of the earth's surface and accounting for about 60% of global freshwater flow. In such basins, the concept of river basin management can be implemented only partial in sub-basins. Nevertheless, IWRM is necessary to observe constraints under international agreements however difficult they are to reach and implement. There are many cases of trans-boundary agreements such as the Colorado, the Nile, the Rhine, the Danube, the Tigris-Euphrates, the Amu Darya and the Syr Darya, the Ganges-Brahmaputra, the Mekong, the Amur Rivers, etc. They have been accumulating a long history of negotiations, some wise and fair but some not necessarily, quite often controlled by upstream nations or strong nations.

4.1.2 IWRM as a UN's guiding principle

The idea of IWRM has been developed and exercised in a long history of managing rivers, lakes and other freshwater bodies for flood management and water utilization since early days of human settlements in Egypt, China,

Indonesia, Japan or any other countries. Especially in agricultural practices, the given land and water resources have been managed to accord with the given nature of hydrological and eco-hydrological cycles and environmental and geographical conditions of the basin, to support necessary socio-economic activities in the basin. In the 1930s, Tennessee Valley Authority (TVA) had developed many multi-purpose dams for navigation, flood control, hydro-electric power generation, together with reforestation, agricultural and industrial development, malaria control, and other economic development of society (Lilienthal, 1944).

The Tennessee Valley Authority Act (TVA Act) was signed by President Franklin Delano Roosevelt on 18 May 1933, as part of his "new deal" policy to recover from the Great Depression occurred in 1929. The act was defined as:

> An act to improve the navigability and to provide for the flood control of the Tennessee River; to provide for reforestation and the proper use of marginal lands in the Tennessee Valley; to provide for the agricultural and industrial development of said valley; to provide for the national defense by the creation of a corporation for the operation of Government properties at and near Muscle Shoals in the State of Alabama, and for other purposes (TVA Act, 1933).

A similar approach was taken in the Rhone River National Company in France (since 1933) too although it covered not a whole basin. The TVA was an ideal example of IWRM which influenced basin management of many basins such as the Colorado, the Missouri and the Fraser Rivers in USA, the Damodar Valley in India (since 1948), the San Francisco River of N.E. Brazil (since 1948), the Gal Oya River (since 1949) and the Mahaweli River (since 1969) in Sri Lanka, etc. (Barrow, 1998).

The concept of IWRM was later adopted by the United Nations as the principal methodology to implement sustainable development of freshwater resources in its action plan Agenda 21 (Chapter 18) agreed at the United Nations Conference on Environment and Development (UNCED), Rio de Janeiro in June 1992. In fact, IWRM was the only methodology presented in Agenda 21 to implement the new concept, sustainable development for freshwater resources.

Dublin principles

Adoption of IWRM in Agenda 21 was based on the stakeholders' gathering at the International Conference on Water and the Environment (ICWE) in Dublin, Ireland, on 26–31 January 1992, six months before the Rio Summit. They discussed about the methodology of sustainable development for freshwater resources and the Dublin Statement was declared (ICWE, 1992).

The statement includes the following four guiding principles, often called as the Dublin principles:

Principle No. 1. Fresh water is a finite and vulnerable resource, essential to sustain life, development and the environment.

Since water sustains life, effective management of water resources demands a holistic approach, linking social and economic development with protection of natural ecosystems. Effective management links land and water uses across the whole of a catchment area or groundwater aquifer.

Principle No. 2. Water development and management should be based on a participatory approach, involving users, planners and policymakers at all levels.

The participatory approach involves raising awareness of the importance of water among policymakers and the general public. It means that decisions are taken at the lowest appropriate level, with full public consultation and involvement of users in the planning and implementation of water projects.

Principle No. 3. Women play a central part in the provision, management and safeguarding of water.

This pivotal role of women as providers and users of water and guardians of the living environment has seldom been reflected in institutional arrangements for the development and management of water resources. Acceptance and implementation of this principle requires positive policies to address women's specific needs and to equip and empower women to participate at all levels in water resources programmes, including decision-making and implementation, in ways defined by them.

Principle No. 4. Water has an economic value in all its competing uses and should be recognized as an economic good.

Within this principle, it is vital to recognize first the basic right of all human beings to have access to clean water and sanitation at an affordable price. Past failure to recognize the economic value of water has led to wasteful and environmentally damaging uses of the resource. Managing water as an economic good is an important way of achieving efficient and equitable use, and of encouraging conservation and protection of water resources.

The first principle directly requires integrated water resources management as a holistic approach to achieve sustainable development, and others indicate the important rules to follow when integrated water resources management is to be exercised, that is, participatory approach, role of women and economic good.

Participatory approach including women has been recognized increasingly important in the MDGs, SDGs and HFA and SF as seen in the previous chapter. Role of women was in principle 20 of the Rio Declaration on Environment and Development, saying "Women have a vital role in environmental management and development. Their full participation is therefore essential to achieve sustainable development" (UNGA, 1992). The IWRM Dublin Principle played an important role to put it in the declaration. It reflects that in domestic matters, especially in water management, women play a critical role and without well institutionalized their involvement and empowerment scheme, no sound water management nor societal development is possible.

Another impact that this principle made to the later world was regarding the fourth principle "water should be recognized as an economic good". This statement is meant for promotion of efficient use of water and proper management and provision of water for those who suffer from safe drinking water and basic sanitation. It encouraged the private sector to get involved in water business and, at the same time, created concerns on a conflict of interest between the public provision of water and the corporate interest of private management. In a number of ministerial conferences held afterwards such as on the occasions of World Water Forums in Hague in 2000, Kyoto in 2003 organized by the World Water Council (WWC) established in 1996 (WWC, 2004), etc., the Dublin principle was referred in their declarations and privatization was supported under the idea of public–private partnership (PPP). Against such movement, a strong anti-privatization movement was developed claiming that water is a public good and a human right. It was often joined by the anti-dam movement against environmental degradation and relocation of people, and their demonstration became sometimes extreme. The debates have still been continuing (Bakker, 2007; Fletcher, 2018). One of the issues was how much water is a human right. For this, Peter Gleick, a US water resources scientist, proposed 50 liters per day per person consisting of drinking water 5, sanitation services 20, bathing 15 and food preparation 10 liters (Gleick, 1998).

IWRM in Agenda 21

Agenda 21 Chapter 18 "Protection of the quality and supply of freshwater resources: application of integrated approaches to the development, management and use of water resources" says, at the beginning of program area: Integrated water resources development and management,

> 18.6. . . . The holistic management of freshwater as a finite and vulnerable resource, and the integration of sectoral water plans and programmes within the framework of national economic and social policy, are of paramount importance for action in the 1990s and beyond. The fragmentation of responsibilities for water resources development among sectoral agencies is proving, however, to be an even greater impediment

to promoting integrated water management than had been anticipated. Effective implementation and coordination mechanisms are required.

This is the statement of the need of integrated approach, to realize holistic management, fragmented sectoral agencies are impediments. Then about IWRM, Agenda 21 further states that:

> 18.8. Integrated water resources management is based on the perception of water as an integral part of the ecosystem, a natural resource and a social and economic good, whose quantity and quality determine the nature of its utilization. . . . In developing and using water resources, priority has to be given to the satisfaction of basic needs and the safeguarding of ecosystems. Beyond these requirements, however, water users should be charged appropriately.

This reflects the Dublin Principles saying "water as an integral part of the ecosystem". Also, "should be charged appropriately" but added only "beyond basic needs and the safeguarding of ecosystems". It continues:

> 18.9. Integrated water resources management, including the integration of land- and water-related aspects, should be carried out at the level of the catchment basin or sub-basin. Four principal objectives should be pursued, as follows:
>
> a. To promote a dynamic, interactive, iterative and multisectoral approach . . . that integrates technological, socio-economic, environmental and human health considerations;
> b. To plan . . . based on community needs and priorities within the framework of national economic development policy;
> c. To design, implement and evaluate projects . . . based on an approach of full public participation, including that of women, youth, indigenous people and local communities in water management policy-making and decision-making;
> d. . . . to ensure that water policy and its implementation are a catalyst for sustainable social progress and economic growth.

Public multi-sectoral approach, community needs and priorities within national policy, participatory approach, catalyst for societal progress are all mentioned.

4.1.3 The conceptualization of IWRM

Nevertheless, it is still vague and need further efforts to translate the concept into a more concrete form. To follow are some example efforts to make it operational.

What to integrate for IWRM?

IWRM calls upon integration of various water-related factors and activities. What factors and activities to be integrated? Figure 4.2 tries to illustrate the complex nature of components to be integrated. They are first classified into three categories: nature and hydrological cycle, land and water users and socio-economic system. Then together with their intersections, there are seven areas of classified components. But any target geographical area of IWRM is not independent from the outer world. It receives external impacts and interacts with such as climate change, migration, etc. In the form of products, water can be exported from a watershed to another and from a nation to another by trade as "virtual water". IWRM lies at the intersection of all those factors.

The three main areas represent the following factors, components and players which affect the integration process of water resources management.

- *Nature and hydrological cycle* (left circle): It is the water-related nature given to human beings, namely, geology, topography, climate, precipitation, evapo-transpiration, infiltration, soil-moisture, surface-discharge, groundwater and water quality. River water produces sediments. Water availability is probabilistic and unevenly distributed in time and space which causes extreme events such as heavy rains, floods, droughts and debris flows. Human beings are living with such nature and hydrological cycle and they set natural conditions of water management. The hydrological cycle in a basin or a nation is affected by external inputs, too such as climate change, inter-basin water transfer and external pollutants, where climate change is seriously impacting basin hydrology especially extremes in recent years.
- *Land and water users* (right circle): Land and water users are the major stakeholders of water management. It starts from eco-system, then to human side, the largest user is agriculture followed by industry and urban activities (Shiklomanov, 1998). Other than those water supply, there are other major users, namely, hydropower generation, navigation, recreation and aesthetic landscapes. They determine physical targets of necessary actions of water management. Water users change with migration in and out which impacts the basis of stakeholders. Water use in a basin is connected with outer world in the form of foods or other products as "virtual water" as well as by supply chain, too. Water users' community is much affected by such interactions.
- *Socio-economic structure* (bottom circle): The organizations responsible to various parts of socio-economic structure are also important stakeholders. Water resources are administered by governments of various levels where administrators, legislators and judicial courts are involved. There are politicians, various NGOs, research institutes for science and technology, media, education sector, cultural sector, etc. are important stakeholders in various aspects of decision-making. International assistance programs affect the domestic decision-making and budgets.

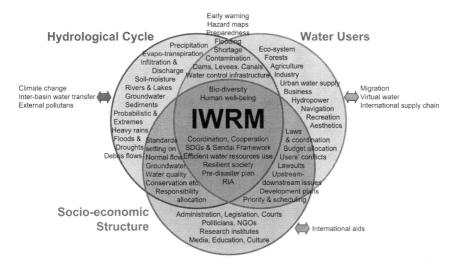

Figure 4.2 Complex considerations of components to be integrated in IWRM planning. IWRM integration targets are classified into nature and hydrological cycle, water and land users, socio-economic structure and their interactions which are subject to outer influences such as climate change and migration indicated by dual direction arrows. IWRM targets are in the center, human well-being and biodiversity through sustainable development.

In the joint areas of those main players have many components to be integrated, too.

Between *hydrological cycle* and *water users*: Flood-related hazards and risk reduction infrastructure such as dams and levees. Early warning belongs to this area.

Between *water users* and *socio-economic structure*: Here all laws and regulations belong. Also users' conflict coordination in budgets, storage, priority, etc. which may lead to lawsuits.

Between *socio-economic structure* and *hydrological cycle*: This area includes standards setting such as on normal flow, groundwater levels, water quality, river classification, etc. Also, the responsibility allocation through mandates among governmental agencies and offices are important factors to be considered in integration process.

IWRM: Finally all those factors are integrated, it is IWRM in the center where by good integration and coordination, water resource systems are well functioning to realize human well-being and biodiversity at the end. SDGs and Sendai Framework will be achieved and resilient society is built through pre-disaster plans, risk impact assessment (RIA), etc.

GWP definition of IWRM

Global Water Partnership (GWP) was an international NGO established in 1996 to foster IWRM employed in Agenda 21 to achieve sustainable development of freshwater resources especially in developing countries.

Originally it was supported by the World Bank, the United Nations Development Program (UNDP) and the Swedish International Development Cooperation Agency (SIDA). GWP is now a network of over 3000 partners organizations in 179 countries such as governmental, non-governmental, academic or private institutes, international organizations including UN agencies, professional associations and any other entities except individuals (GWP: website).

Their main activities are capacity building and knowledge sharing on IWRM and proposed a definition of IWRM which is now most widely referred in water community. Namely, GWP Technical Advisory Committee (TAC) Background Papers No. 4, 2000 (GWP, 2000) proposed the following definition:

> IWRM is a process which promotes the coordinated development and management of water, land and related resources, in order to maximize the resultant economic and social welfare in an equitable manner without compromising the sustainability of vital ecosystems.

This definition says that the essence of IWRM is in a "process of coordination". Water management always involves a series of decision-making at various levels and occasions. Therefore the definition indicates an importance of coordinated decision-making among all related stakeholders. In case of a levee construction, for example, the river bureau of the government would assume the responsibility and initiate a plan with in-house engineers and necessary out-sourcing, consulting with local communities, private and academic experts, environment and ecology sector, budgetary office, etc. considering such as local life, business, industry, infrastructure including irrigation, fishery, water supply, sewerage, hydro-power, road and transportation, other utilities, etc. The success of IWRM policy would depends on how far such related factors are well considered and integrated in the design, assessment, selection and implementation procedure of the plan. In reality, there are quite a few examples that a levee was constructed without discussing and coordinating with the road sector with respect to bridges or conduits to cross each other, and vice versa, a road was constructed without consulting with the flood sector, resulting in low disaster safety. Indeed, if there is lack of coordination with related stakeholders, it is not IWRM.

4.1.4 UNESCO guideline "IWRM at river basin level"

UNESCO published a guideline of "IWRM at river basin level" in 2009 (UNESCO, 2009) as a contribution to the United Nations World Water Assessment Program (WWAP) that was established in 2000. This guideline describes a concrete operational approach for IWRM by which anybody can learn and implement IWRM at river basin level.

The guideline consists of five volumes: principles, coordination, flood, irrigation and environment. Namely, "Part 1: Principles" provides basic principles of IWRM, mainly targeting policymakers, and explains benefits of IWRM at river basin level and the need of promoting it at the policy level. It also proposes a spiral model of IWRM, which illustrates the step-by-step evolving nature of the IWRM process. "Part 2.1: Guidelines for IWRM Coordination" is for practitioners involved in IWRM coordination. It can be used as an introductory guidance for those who are tackling IWRM for the first time, or as a training material for intermediary practitioners and trainers of IWRM.

Then for IWRM practitioners in three thematic areas: "Part 2.2: Flood Management", "Part 2.3: Invitation to IWRM for Irrigation Practitioners" and "Part 2.4: Guidelines for Managing Environmental Sustainability".

The purpose of publishing these guidelines was to raise awareness of the need for an IWRM approach among policy- and decision-makers to improve water resource systems and to facilitate practical end-users with the implementation procedure of IWRM at river basin level. To follow are *the author's interpretation* of the guideline with author's emphases.

UNESCO definition of IWRM

The guideline starts from definition of IWRM as follows:

> Integrated water resources management (IWRM) is a step-by-step process of managing water resources in a harmonious and environmentally sustainable way by gradually uniting stakeholders and involving them in planning and decision-making processes, while accounting for evolving social demands due to such changes as population growth, rising demand for environmental conservation, changes in perspectives of the cultural and economic value of water, and climate change. It is an open-ended process that evolves in a spiral manner over time as one moves towards more coordinated water resources management.

This definition describes the process of coordination more clearly than the GWP's. It says that the process is "gradually uniting stakeholders" and "involving them in planning and decision-making processes". Further, it says "it is an open-ended process that evolves in a spiral manner over time". They consider that IWRM is not an easy matter but can go only gradually, step by step, yet continuously over time. It is a pragmatic approach easily acceptable by a majority of organizations. The meaning of "spiral manner" will be closely looked at following the next section.

Principles in the guideline

Part 1 of the guideline lists overarching principles of IWRM which may be summarized as follows:

1. Water is necessary for sustainable development (all lives, development and environment) so that integrated management is necessary and IWRM is beneficial for all.
2. Objective of IWRM is to attain water security.
3. River basin is an appropriate unit for IWRM. IRBM is IWRM at river basin level.
4. IWRM progress is an evolutional "spiral process". Evolution may be triggered by disasters, crisis, change of political leadership, socio-economic development, environmental conditions, technical innovation etc.
5. The need of IWRM should be understood by top politicians, participated by all stakeholders including bottom-up process in transparent and accountable manner.
6. All stakeholders participating to coordinating process for integration should be aware of and faithful to their own mandates and responsibilities in a clearly defined administrative and institutional structure of a society.

There are some definitions on water security. Among them, the following by African Ministers' Council on Water (AMCOW) and the African Development Bank (AfDB) at Tunis, 28 January 2008 would be reasonable and the simplest, that is:

> Water security is the capacity to provide sufficient and sustainable quantity and quality of water for all types of water services and protect society and the environment from water-related disasters.

It is the state of water being sufficiently and sustainably secured in good quality and quantity free from water-related disasters.

Spiral process

The core concept of the guideline is a "spiral process" as depicted in Figure 4.3, which illustrates the evolving and dynamic nature of the IWRM process. This is the principal strategy of the guideline for implementing IWRA. It claims that the IWRM is not a single shot action but an "evolving and dynamic" process of IWRM towards further "progress of IWRM", each round initiated by some occasions of social and/or environmental impacts.

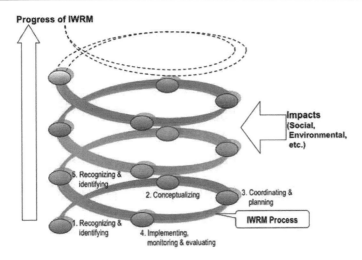

Figure 4.3 IWRM spiral and process. After UNESCO (2009).

As shown in Figure 4.3, the one spiral round of the IWRM process includes the following four steps:

(1) "Recognizing and identifying" the pressing issues of the society and needs of IWRM,
(2) "Conceptualizing" the problem and formulating possible solutions,
(3) "Coordinating and planning" among stakeholders in order to reach an agreement, and
(4) "Implementing, monitoring and evaluating" the plan and its outcomes.

In step (4), during the implementation stage, the plan decided in step (3) may become necessary to be slightly modified or adjusted and fixed within the round. But a major revision should wait until the next round when some societal needs eventually increase and become pressing for further round of improvement of IWRM. Another round can start usually only with some triggering event that gives a political rationale on "why now?". Such an occasion is often created by some disasters, intolerable problems or widely supported societal challenges such as water pollution and water shortages due to urbanization, industrialization, etc., increasing societal needs for better environment and change in lifestyle, inviting Olympic games, and most often new administration with new political leadership.

The core concept of this spiral approach is a recognition of different societies being in different levels of IWRM stages. It is not only level of infrastructure facilitated but also institutional level of resource management, coordination and integration capacity and socio-economic development stage depending on different societal history. There is no universal standard for IWRM but depends on the socio-economic-cultural conditions of society.

Four steps in a spiral round

The suggested four steps: recognizing and identifying, conceptualizing, coordinating and planning, and implementing, monitoring and evaluating are the concrete actions involved in this approach, IWRM at river basin level.

Before getting into detail steps, it is necessary to identify who initiates the IWRM action. It is in general a governmental agency or office which has responsibility and mandate on the concerned matter. In fact, the organization which initiated the action should take responsibility to lead the matter throughout the entire process of IWRM although sometimes the leadership eventually becomes a joint one with other agencies. In any nation, there is a historical organizational structure, and each agency in the administrative structure has its own mandate. From time to time, such a structure may change by political decisions to seek better administrative efficiency, but any given administrative allocation of mandates is an allocation of responsibility with budget and human resources. Therefore, a responsible bureau, agency or office assumes responsibility to initiate, do fact finding, make proposals, conduct coordination for agreement and finally implement the plan. It looks like a sectoral approach. Yes, it is so as long as responsible leadership is concerned. Of course, any others can express their opinion and appeal what to do, and the responsible agency has to carefully listen to their voices, but no decision or implementation is possible without official mandates. Involvement of all stakeholders in an equal basis in the first place is an impractical approach that may create difficulty and complication. Sometimes strong NGOs or even UN aid agencies may step in implementation stage such as during recovery from catastrophic disasters but still it should be together with and in accordance with national responsible agencies. Otherwise, an eventual confusion may be inevitable.

Now detailed steps:

Steps (1) Recognizing and identifying

The first step of an IWRM cycle is to "recognize the need for IWRM through identification of needs and problems in the basin". Naturally, it should start from identifying issues and problems that a society is facing. This is the fact-finding step. But what is special to the guideline is that through the identification of problems, the managers are instructed to recognize the need for

IWRM to improve the situation. If a mayor faces a water shortage problem in his/her city, the guideline suggests not only to find out the physical balance of water demand and supply, and the critical paths to solve the problem under given conditions of demographic, socio-economic, environmental, etc. but also look at the total water system and consider if all stakeholders are well-participated, coordinated and settled in good agreement and whether the definition of IWRM is realized or not. The mayor has to realize that if any room is left, that room is the target of the new IWRM improvement cycle.

Step (2) Conceptualization

This is the step of conceptualizing a solution considering societal needs, constraints and balance with other stakeholders' plans and concerns. This step starts from analyzing and assessing the basic findings in the 1st step and making further hearings towards solution finding. In hearings, it is important to include stakeholders outside of water sector to seek harmony with related plans and activities such as regional development, land use planning and disaster management. Hearing from other stakeholders in this conceptualization step is already the start of coordination process, without which, in the 3rd step for final agreement, many time-consuming revisions may become necessary and, in the worst case, even overturn may happen. Other basic needs for conceptualizing a solution would include careful consideration of priorities with respect to timing and budget, anticipated future changes in population, urbanization, industrialization and other socio-economic changes, environmental and climate changes, technologies both advanced and conventional, etc. It is also necessary to list up all the related stakeholders and their concerns for the next step. Preferably, alternative plans and options are prepared to show stakeholders in coordination for the final agreement.

Step (3) Coordinating and planning

This is the step to finalize the solution conceptualized in the 2nd step into a detailed plan with all stakeholders' agreement. Key processes of this step are as follows: All relevant stakeholders have to participate. Specifically, the environmental sector's involvement is necessary and if no representative is available, the coordinator has to assume to be its spokesperson. It is important to select a lead coordinator (organization or individual) who is supported by majority of participating stakeholders. It is quite often the case that the responsible organization plays by itself or select a person for a coordinator as they are most competent and knowledgeable in the subject area. Involvement of stakeholders should be timely to avoid unnecessary duplication of discussions. Stakeholders' participation to the process should be in an equal basis with clear understanding of the problems and situations, supported by transparent full disclosure of information by the responsible agency and participants. The coordination process has to aim to reach consensus including cost allocation among stakeholders at final stage. Modification by coordination is better to be favorable for and equitable among all stakeholders (seeking the Pareto optimum). Agreement is necessary for the cases of emergency with extreme occasions. If there is no sign of reaching agreement in the process, it should go back to the previous step and restart conceptualizing again.

For effective coordination, it is important to have a well-defined administrative system established in a society, with sectoral allocation of mandates and responsibility covering all subject areas exhaustibly without being left out at all levels, national, regional and local. In other words, all participants should have clear area of interest and responsibility in relation to the proposed plan. Nobody who does not have direct problems to share and responsibility to contribute should join the stakeholders' group to be coordinated.

Note that this does not exclude the non-governmental organizations or civic society who have no direct interests as local residents but have views to express as experts or opinion leaders. Their role is important and has a right to be involved in discussion but their responsibility should also be clear that they present credible facts and justifiable opinions, and follow the rule set by the coordinator.

In order to make this possible, all sectors to be coordinated have to have a full access to related data and information. Conversely, all sectors to be coordinated have the responsibility to understand the situation, follow the rule and play their role if agreed. In other words, transparency of information and shared responsibility on the project are mandatory in participating in coordination process for integration, which is a rule of participatory approach, too.

Such coordination process often takes place by administrative offices circulating proposals for public hearing for a specified period. Interested stakeholders are requested to submit their views in a written form which will be considered by the responsible agency and eventually give answers back to the stakeholder. Public hearing for dialogue in a large hall is also a typical form that goes along the circulation of written documents.

Step (4) Implementing, monitoring and evaluating

The final step is implementation, monitoring and evaluation. The plan of project such as infrastructure development or institutional framework establishment should be promptly executed, developed and put to use for operation. The system should function well. Such the process of implementation should be monitored, and the outcomes and impacts should be evaluated.

For this, the network of stakeholders participated in the coordination step for the plan making should be maintained and well informed which serve for efficient adjustment of the system for unexpected problems realized as well as balanced evaluation of the outcomes. The initial stage adjustment of a system is quite important as in general no system functions perfect as planned from the beginning and some adjustment is necessary. The data collected are useful information for the next round of IWRM process as well as similar IWRM projects elsewhere.

The volumes of Part 2.2–2.4 of the guideline on flood, irrigation and environmental sustainability include various cases of respective themes in different countries. They are customization of the process to particular cases.

4.2 INTEGRATED FLOOD RISK MANAGEMENT (IFRM)

In the previous section, IWRM process was discussed in detail. In this section, IFRM, the main theme of this textbook will be presented. IFRM is part of IWRM as repeatedly mentioned and the discussion on IFRM is only a focus of IWRM on flood risk. It is therefore not at all independent from IWRM discussion but should be considered in its part. The discussion in this section is brief as the dense case studies in Japan follow in Chapter 5, especially in Sections 5.4–5.6.

4.2.1 Introduction

This section first introduces definitions of flood, flood risk and integrated flood risk management (IFRM) and, in its relation, activities of the European Flood Directive. Next, the difference between flood risk management and flood control is explained and finally a brief overview of IFRM is presented.

Flood-related definitions and the European Flood Directive

Definitions of floods are different for different purposes. Japan Meteorological Agency (JMA), for example, defines that (JMA, Home Page on Terminology):

> Flood is overflow from normal riverbed to flood terrace in a river or overflow from banks to outer area of floodplain by abnormal increase of discharge or water level of a river.

Also it says that:

> In hydrological terminology, flood is an abnormal increase of water level or discharge in a river by rainfall or snowmelt.

Their focus is basically rivers and riverine floods, concerning whether "overflow from normal riverbed to flood terrace in a river or overflow from banks to outer area" occurs or not. This focus is quite reasonable as riverine floods are most concerned and decisive to society. However, they do not include inland floods, urban floods, sewer floods, coastal floods, etc. which are also quite influential in urban and coastal regions.

European Flood Directive defines flood in its Article 2 as follows (European Flood Directive, 2007):

> "flood" means the temporary covering by water of land not normally covered by water. This shall include floods from rivers, mountain torrents, Mediterranean ephemeral water courses, and floods from the sea in coastal areas, and may exclude floods from sewerage systems.

The first sentence "'flood' means the temporary covering by water of land not normally covered by water" is a simple but precise definition of flood which may be quite practical to use. In fact, it must be the most often cited definition in recent literature. But in the latter part, this definition excludes floods from sewerage system including storm water drainage which are critical components of urban floods and inland floods. The omission of such an important category of floods would be due to administrative reasons of the Directive.

There is another definition of flood, that is, the IPCC Special Report on Managing the Risks of Extreme Events and Disasters to Advance Climate Change Adaptation (SREX) defined flood as follows (IPCC, 2012):

> The overflowing of the normal confines of a stream or other body of water, or the accumulation of water over areas that are not normally submerged. Floods include river (fluvial) floods, flash floods, urban floods, pluvial floods, sewer floods, coastal floods, and glacial lake outburst floods.

The core concept of this definition includes JMA's, saying "overflowing of normal confine of a stream" and the EU's, saying "water over the area not normally submerged". As this definition is purely scientific and includes almost all aspects of other definitions, this textbook considers it most appropriate for a universal use.

About *flood risk*, the EU flood Directive Article 2 defines it as follows:

> "flood risk" means the combination of the probability of a flood event and of the potential adverse consequences for human health, the environment, cultural heritage and economic activity associated with a flood event.

This is a combination of UNISDR definitions of *risk* and *disaster risk* shown in Section 2.1.2 into *flood risk*. Although it does not say, it is an expected loss discussed in Section 2.1.3 and the same formula Equation (2.4) applies for estimation.

Now definition of integrated flood risk management (IFRM). It is defined as:

> IFRM is a process which promotes the coordinated management of flood risk as part of IWRM to minimize the negative impacts of floods by developing infrastructure, reducing exposure and vulnerability, and increasing preparedness of society while maximizing benefits of floods on human well-being and biodiversity.

This is an extension of IWRM definition to focus on flood risk. It emphasizes coordination process among all related matters. It is not just coordination between roads and levees or public sectors and communities but among all related components and units including biodiversity in societal system.

In this line of thoughts, the definition of "integrated flood management" is similar. It is just a reverse of a focus to "maximize the benefit of floods":

> Integrated flood management (IFM) is a process which promotes the coordinated management of floods as part of IWRM to maximize the benefit of floods while minimizing the negative impacts of floods on human well-being and biodiversity.

Floods are not necessarily hazardous matters but basically precious resources for many nations as seen in the Nile and other cradles of civilization for agriculture. Especially in South, Southeast and East Asia where the main diet is rice, seasonal floods have been indispensable resources for rice production. Accordingly, rice cultivation has been the main control of societal settlements: flood management system, development of residential areas, transportation routes, and all other livelihood settings that have been developed in the way most opted for rice production in a basin.

In fact, there is no flood risk management if beneficial aspects of floods are neglected. Also, all terminology related to water resources management becomes necessarily similar as long as it is an "integrated" management, because by integrating all the water-related components in society, eventually all matters of society come into a picture and any key solutions necessarily become part of the total societal solution.

Here, the European Flood Directive is briefly introduced as it is important not only with its good definition on flood but also with their challenging international cooperation for a large part of Europe. It is an integral part of its preceding agreement, EU Water Framework Directive (WFD) adopted by the European Parliament on 23 October 2000. The WFD responded to the necessity of an integrated river basin management in Europe especially on water quality issues of rivers and water quality and over-pumping of groundwater. One of the triggering incidents for this agreement was a catastrophic chemical contamination accident of the Rhine River that occurred on 1 November 1986 by a storehouse fire near Basel, Switzerland. With the firefighting water, a huge amount of chemical substance was discharged into the Rhine and damaged downstream countries a great deal (Giger, 2009). Since Europe shares transboundary rivers and groundwater, their collaboration for integrated water management was imperative (European Commission: website on EU Water Framework Directive).

The Flood Directive is an agreement made on 23 October 2007 among all the EU countries to observe on flood management. It takes a six-year cyclic operation which is coordinated and synchronized with the WFD implementation cycle. The current Flood Directive Article 14 says:

1. The preliminary flood risk assessment, or the assessment and decisions referred to in Article 13 (1), shall be reviewed, and if necessary updated, by 22 December 2018 and every six years thereafter.
2. The flood hazard maps and the flood risk maps shall be reviewed, and if necessary updated, by 22 December 2019 and every six years thereafter.
3. The flood risk management plan(s) shall be reviewed, and if necessary updated, including the components set out in part B of the Annex, by 22 December 2021 and every six years thereafter.

Also, Article 16 says:

The Commission shall, by 22 December 2018, and every six years thereafter, submit to the European Parliament and to the Council a report on the implementation of this Directive. The impact of climate change shall be taken into account in drawing up this report.

Thus, all the member countries are requested to repeat all the processes of assessment, hazard/risk maps review, management plan every 6 years and their implementation report to the European parliament also every 6 years. This is a great achievement of regional cooperation in the European Union.

Flood risk management and flood control

Flood risk management is different from flood control or flood defense. Flood control tries to stop floods to happen or reduce their scale by engineering infrastructure such as dams, levees, diversion canals, retardation ponds, underground rivers, etc. While those engineering structures are important and indispensable to protect areas exposed to flood hazards, it is obvious that floods can never be completely controlled by those structures but only up to certain design level for which structures are constructed to resist if no unexpected failures happen. But excess floods or failures do occur; however, a huge investment is made for flood control infrastructure.

Such shift from flood defense to flood risk management is often considered as a recent major policy shift such as since 1990's UNIDNDR efforts in disaster research mentioned in Section 3.2.1 (Schanze, 2006; Butler and Pidgeon, 2011). But in fact, in a long human history, flood defense has been subordinate to risk management as human beings did not have enough power to control extreme nature until recently. Therefore a major strategy has long been adjustment to floods by human side. The flood defense efforts became prevailing after major infrastructure construction became possible, but it was soon realized that complete structural control of floods was impossible, especially with recent climate change.

Flood risk management aims to manage flood risk rather than physical flood hazard itself. In terms of vulnerability, risk management tries to reduce societal vulnerability from all possible means, exposure, basic vulnerability and coping capacity, that is, EV_B/C in terms of Section 2.1.3. Especially in coping capacity C, emphasized are non-structural means and cooperation among related components as compared with structural means to control floods. By doing so, flood risk management can reduce expected losses by flood disasters. This is also expressed as an increase of resilience against flood hazards by increasing power to resist, absorb, accommodate and recover from the effects of a flood.

Nevertheless, those concepts, flood risk management, flood control or defense are not at all mutually exclusive. For flood risk management, flood control facilities are indispensable to prevent especially frequent type, small scale floods. For economic growth, physical protection is indispensable to protect all of human lives, properties and livelihood together. In order to accumulate economic wealth, frequent floods should by all means be physically controlled without necessitating evacuation.

At a glance of methodologies of FRM

Methodologies of FRM are reduction of societal vulnerability against flood hazards basically by reducing exposure to risk area, constructing physical infrastructure, and promoting preparedness especially early waring and evacuation. To follow are their brief descriptions:

- Reduction of exposure from flood risk area is a fundamental means for flood risk management. There are various supporting means such as land use regulation guided by flood hazard maps, subsidies for resettlement to safer places including livelihood, compulsory insurance for building houses in risk area, say, 1/100 year or more frequent floods, and regulation of illegal settlements to avoid squatters. In the future settlement in compact cities would be one of ultimate solutions for exposure control, which is good not only for flood and other hazards management but also protection of biodiversity, energy efficiency, environmental conservation, etc. See Section 6.1.4 for more discussion.

- Reduction of frequency and magnitude of floods is possible either by storing more water in a basin before going down to rivers or by confining water into rivers and more quickly flow it down to the sea. Storing more water in a basin is by such as forestation, dams, sabo works, retardation ponds, in-situ infiltration facilities, open dikes, etc. More quickly to the sea is by such as levees, channel improvements, diversion canals, etc. They are both physical flood control measures, mostly engineering and some forests and slopes management. They cost large investment for construction and OMR (operation, maintenance and replacement/repairment).
- The underling risk should be reduced during development stage of land and structures. Such development as reclamation of wetland, conversion of forests or agricultural fields to residential area or a large-scale infrastructure construction such as highway, industrial complex or underground cities is decisive in creating risk unless proper risk reduction measures are taken. In order to make sure development does not create new risk, pre-assessment is necessary. Namely, impacts on hydrological and other environmental processes and resulting impacts on disaster risk should be identified before development takes place which is called risk impact assessment (RIA) similar to environmental impact assessment (EIA) as mentioned in Section 2.4.1. Based on RIA, proper countermeasures should be taken and until the safety is verified, no development plan should be permitted. In other words, RIA should be exercised as compulsory regulation. Quite often the impact of land use change is not carefully identified before development decision is made and creates new risk which should by all means be avoided as the Goal of Sendai Framework paragraph 17 states "prevent new and reduce existing risk" (UNISDR, 2015).
- Pre-planning is an important strategy both for build back better including exposure regulation and infrastructure construction. Any fundamental societal transformation to safer cities needs a good timing and occasion to initiate, for which the occasion of a tragic disaster should not be waisted as an opportunity window. To make it possible, pre-disaster planning is indispensable as reconstruction starts immediately after the emergency response and rehabilitation is already the start of reconstruction. Pre-planning is important for emergency response, too. Without planning before disaster, no wise operation at emergency is possible.
- Emergency preparedness is necessary to prepare before, during and after the attack of a hazard. Before the arrival of a hazard, it is necessary to make sure maintenance of infrastructure to properly function

and repairment or emergency support ready. Preparedness of rescue, evacuation and gathering information to grasp situation for proper disaster management is also critical to stop expansion of disaster impacts. Emergency response operation during and after the arrival of a hazard is only an exercise of what prepared before arrival of a hazard. In order to make sure what prepared well function, training and exercises in daily life are necessary and regular drills by community residents are critical. Nevertheless, there are always various unexpected things happen. Whenever such happens, all available knowledge and resources should be organized and exercised.

- The most important preparedness for emergency response is early warning, for which observation, data transmission and analyses are necessarily supported by continuous efforts of science and technology.
- For all those, the basis is the support of people based on their awareness of risk, disaster literacy and disaster consciousness. Again, disaster management is a matter of human empowerment necessary for sustainable development.

4.2.2 What and how to integrate for IFRM?

Stakeholders and other factors to be integrated

Integration is emphasized but what and how to integrate for flood management? First, what to integrate? Suppose there is a project plan for flood risk reduction. As all components of a society are one way or another interconnected and interdependent with each other, implementation of any plan would give some impacts to some components of a society, and from those components to other components. This impact propagation forms an impact network that should be the target of integration with the new project plan. It is the same as a supply chain or value chain in production or sales industry. Then how to identify such an impact network? In real society, there are always distinct stakeholders (or sectors) who represent a certain area of interests and share a socio-economic or environmental stand. Thus the important action of IFRM is *integration over all related-stakeholders* in the process of coordination. In the Asian Disaster Preparedness Center (ADCP) report "A primer: Integrated flood risk management in Asia" (ADCP, 2005) the potential stakeholders are listed as follows:

- The emergency planning and response agency or committee
- Government ministries and departments at all levels
- National Disaster Management Office (NDMO)
- Services providers (water, sewerage, highways, transport, public works, energy and telecommunications, etc.)

- Sectors (construction, education, health, industry, agriculture, etc.)
- Universities, research groups and training organizations
- Local voluntary and community organizations
- Flood, storm and earthquake control commissions
- Private/Business sector
- Insurance sector
- International agencies/NGOs
- Armed forces, police and fire service
- Marginalized and minority groups

This list seems incomplete and biased to administrative sectors but would serve as a basis to think of a complete set of stakeholders. Some of the important stakeholders that need to be added would be the groups that concern vulnerable people, gender, ecology, climate change, environment, etc. About vulnerable people to disasters, Karina Vink (2014) made a thorough review of vulnerable people in literature and proposed a metric for measuring the degree of caring vulnerable people in national policy. Her list of potentially vulnerable people (she uses "potentially" as those categories of people are not necessarily always vulnerable) is:

- Women (pregnant women, women with babies, women-headed family)
- Handicapped people (hospitalized people, people in nursery home, mentally retarded people)
- Elderly
- Children
- People living in poverty
- Ethnic minorities

Vink also pointed out other groups such as:

- Foreigners, tourists who are stranger in the area and/or do not understand native language.
- Pet owners who try to provide help for pets as family members.

By comparing policies in the Netherland, Japan and USA, she concluded that "The metric shows that disaster risk management laws rarely anticipate a future increase in the number of potentially vulnerable people, and none of the laws were created by involvement of potentially vulnerable people". It is important to include stakeholders that represent those potentially vulnerable people.

Women in many countries are still put to unequal opportunities and culturally and socially treated subordinate. In order to reform such an unfair society, there are many groups working on gender issues that form an important stakeholder, similar to environmentalists and ecologists for conserving biodiversity.

By coordinating all those stakeholders, IFRM integrates the related administrative, institutional and political units over *all sectors*, from public to private and from academia to communities, and *all levels from local to national*, whose collaboration is critical to flood management. It also integrates over *all means of hazard management*, both structural and nonstructural, and *all phases of disaster management cycle* from emergency responses, recovery, development and preparedness.

Other than stakeholders, integration is necessary also *over time and space*. That is to say, a system design should integrate over time with the system development, operation, maintenance, replacement and finally termination processes over entire lifecycle. The time horizon of a system to assess the performance, for example, in 10 years' or 30 years' operation may give totally different pictures such as in terms of benefit–cost ratio. A system design should also integrate over space, namely, upstream and downstream, riparian and hinterland, flood plain and mountains, surface and underground, one basin to another, etc. In the case of transboundary rivers, integration is especially difficult. Quite often, upstream improvement of flood safety gives adverse impacts to downstream, and conversely, upstream flooding saves downstream. To settle such conflicting issues, technological ingenuities, political negotiations, priority settings or some compensation arrangements may become necessary.

Another integration need would be *uncertainty*. If probability of occurrence or statistical nature of a phenomenon is known such as heavy rains, their occurrence is called a risk and may be embedded in design by probabilistic treatment. But if no statistical nature is known such as socio-economic changes of future demand or preference, climate changes or occurrence of unexpected accidents, they are called uncertainty and cannot be included in design by formal statistical treatment. Rather different criteria such as *robustness or resilience* should be used for system design. In case of a reservoir design, for example, a large redundant size reservoir may be more robust than the one just sized at time of development as a large one performs well even in case that some new water demand emerges, or some unexpected scale climatic change occurs. If nothing happens or even if demand decreases, it still functions as necessary. So-called freeboard or extra margin of dike height is the same. There are many cases where the facilities work perfectly as long as design criteria are as assumed but easily become useless or even harmful if the societal conditions or preference unexpectedly changes. Selection of location and type of facilities is so important to be robust to unexpected changes of the social demand. If a system is designed too sharp for its particular objective, a slight change of natural or social conditions makes it useless.

To integrate all those factors for IFRM, that is, human stakeholders, systems factors as time, space, uncertainty, robustness, etc., and natural conditions especially hydrological cycle, it is necessary to take a systematic approach with scientific decision-making by all related stakeholders working together. It is a matter of science and technology and a process of collaboration.

Related discussion is in Paul Samuels et al. (2010) report on a framework on IFRM which is a research result of the Integrated Project FLOODsite (2004–2009) funded by the European Commission to provide support for the EU Flood Directive (2007). In conclusion, they say that IFRM is a reflection of policy change from flood defense to flood risk management.

> . . . further evolution of policy and practice is needed to achieve Integrated Flood Risk Management which will support sustainable development by providing a balance between risks, the whole cost and impact of mitigation, economic development and the functioning of vital ecosystems. Clearly this is a demanding aim which will require communication, understanding and cooperation between many technical and professional groups as well as with the broader public.

These components are all to be integrated in IFRM.

About the systematic approach, the next section briefly touches upon and the partnership will be discussed in Section 6.3 under the title *transdisciplinary approach* (TDA) where all disciplines, that is, natural sciences, social sciences, humanities, and all sectors, i.e., academia, public sector, private sector and civic society work together for scientific decision-making and implementation. This is the approach necessary to go beyond the limit of disciplinary and sectoral approach and to make societal transformation possible.

Systems approach for integration

Next, how to integrate? Society is too complex system with many interconnected and interdependent components. Each stakeholder is a subsystem of a total societal system and each subsystem has its own objective, functional mechanism, performance criteria and given constraints. Subsystems and components follow interaction rules, too. IFRM is also a subsystem of a total societal system which has an objective to minimize flood risk while maximizing benefit of floodplain. It should work within a total societal system finding acceptable solutions in accordance or compromise with objectives and constraints of all other subsystems and components. This is a systems approach where selected subsystems and components are managed not exclusively from others but jointly with all related subsystems and components. Detail systems analysis is out of the scope of this textbook but the following rather classic framework on optimization procedure and a fair cost allocation scheme widely exercised are briefly touched upon.

Optimization procedure for a simplified system

Finding acceptable solutions of such a complex system as a whole is too difficult and impractical. For any problem solutions, some simplification is necessary. That is to identify a decision space for a particular sub-system to freely choose treating rest of subsystems as constraints. This is not an easy task but quite often the governing law of an organization identifies such a space in terms of its mandate and responsibility. Under such simplification, it is possible to find an acceptable solution in its decision space. There is a standard procedure to proceed such system optimization process.

Each subsystem has its own objective to achieve, that is, a set of targets to meet, a set of performance criteria to satisfy or some combination of them within a set of constraints given. In IFRM, the objective is to minimize flood disaster risk subject to a given set of performance criteria to meet constraints such as budgets and a given level of benefits to floodplain. System design is to identify a set of components that can achieve the objective within constraints. Looking at a system in such a way, new insights would be found on the function of the system and the role of components or constraints. In order to transform a real system into such an optimization problem, however, much simplification would be necessary and quite often end up with something far from reality. In such a case, its use is limited.

In water resources management, one of the earliest systems analysis presented was by the Harvard Water Program 1955–1965 through its publication "Design of Water-Resource Systems" (Maass et al., 1962) which served like a Bible of system design for some decades. In this book methodology of systems analysis was developed and applied for optimal design of a system of multi-purpose multi-unit dam reservoirs for flood control, irrigation, hydropower, municipal water supply, etc. They considered the size and location of dams as *decision variables* to *maximize* the economic return from the reservoir system (*objective function*). In order to evaluate the system performance for different dam sites and reservoir sizes with different operation rules under different time series of streamflow, they used a *simulation* method to calculate the objective function for all possible combinations of decision variables and obtained a huge number of simulation results. From such simulation results, the optimal design (a set of decision variables to realize the maximum economic return) was selected according to the preset *evaluation criteria* such as a net benefit function or a certain probability of system failure in meeting a target water demand.

Since a long historical streamflow observation data are not always available (especially in 60 years ago), they proposed a method of random data generation by Monte Carlo simulation to get a statistically equally likely streamflow series (*streamflow synthesis*). Using such synthetic streamflow series, stochastic nature of system performance was simulated and assessed. In fact, their idea of streamflow synthesis initiated a new research area of "synthetic hydrology" that lasted a few decades.

Harvard Water Program presented four steps of system design: identifying the objectives of design, translating objectives to design criteria, designing plans to fulfill design criteria, and evaluating the consequences of the plans. These steps must be quite useful in any system design as well as planning process including IFRM. To follow are a modified version of the procedure to make it operational for IFRM:

- Fact finding and problem identification.

  In a concerned region, find population, demography, industrial activities, hydrology, environmental quality, development potentials, etc. and identify needs of flood control, irrigation, urban water supply, poverty alleviation in a certain district, hydropower generation, environmental protection, etc.

- Objectives setting.

  For example: Set goals for economic development including flood damage reduction and conservation of certain environmental quality in the region. Choose methodology by constructing multi-unit multi-purpose dam-reservoirs and levees. Set the objective of the system design as: *Maximize* net benefit or cost-effectiveness of constructing and using multi-unit multi-purpose dam-reservoirs and levees in given potential developmental sites *subject to* a given budget and given target (minimum) levels of irrigation, urban water supply, hydropower generation, flood control and environmental quality to satisfy.

- Design criteria.

  For example: Potential dam sites. Maximum size of dams at each site. Target irrigation (such as in m^3 in specified months). Target urban water supply. Target design flood at a given site. Budget within \$K. Target fish species to be protected. Functional relation between a flood control capacity in a reservoir and necessary height of levees at given sites and length. Hydropower generation is treated as a sub-product. Some set of reservoir operation rules, etc.

- Evaluation criteria.

  For example: Benefit functions of irrigation, urban water supply, flood control and hydropower generation. Cost of dam construction by unit storage volume (non-linear function). Economic loss (or penalty) of failure to meet respective target or weighing factors between *economic benefits* and rate of *system failure* (negative factor) in meeting target supply or flood protection. Robustness of a system or resilience factor may also be included. Evaluation time length (*time horizon*) which is usually a durable life of a dam. Time *discount rate* to translate future cost and benefit to present value.

- Design alternatives.

  Design alternative plans of water resource systems to meet the design criteria. In case of flood management, options of levees, storage reservoirs, replacement of residents, location and size of facilities etc. should be considered. Also for each, options of engineering design on structure, their shapes, combinations, services for multi-purposes etc. should be considered.

- Evaluation of alterative systems or courses of actions.

  Evaluate system performance and consequences of different options under different hydrological series. Here simulation method is suggested to evaluate performance. There are other methods such as mathematical programming including linear programs (LP) and dynamic programs (DP) but their applicability is narrower than simulation in real world.

- Select options according to evaluation criteria.
- Implement the option selected.
- Review consequences (post-audit).
- Adjust the system to accommodate the unexpected performance or consequences.

Note that in this procedure, an objective is set as: maximize (or minimize) an objective function subject to constrains. Quite often any objective can be set not to be maximized (or minimized) but rather to achieve a target level to be satisfied, which is expressed as a constraint.

Note also that there is a similar planning process called PDCA (plan-do-check-action) cycle. This is useful for continuous improvement of a process such as of business operation or manufacturing products that are easily adjustable. But it is not suited for large infrastructure development which is in most cases irreversible and not easily adjustable once it is decided and implemented.

Cost allocation: Separable cost – alternative justifiable-expenditure method

Integration requires a fair rule for all players at all aspects of joint work. It should be throughout the work process including information exchange, designing a plan, negotiation of share of responsibility and all other steps relating to decision-making. Especially in the cost allocation process, fairness is critical and the procedure should be clear and agreeable for all members. In multipurpose-dam construction project, the "Separable cost – alternative justifiable-expenditure method" is specified by the Specific Multipurpose Dam Law Implementation Directorate of 1957 in Japan (Showa 32 Governmental Ordinance 188, 1957) which is a combination of "Separable cost remaining benefit method" and "Separable cost alternative justifiable-expenditure method" described in "Water-Resources Engineering" of Linsley and Franzini (1979). Table 4.1 is the cost allocation procedure exercised in Japan described using a hypothetical dam construction case for multiple purposes, flood mitigation, irrigation and hydropower generation.

It describes the following procedure: Suppose a multi-purpose dam project P is considered to be constructed jointly by users A, B and C. How to allocate the construction cost among those three users? The separable

Table 4.1 Separable Cost – Alternative Justifiable-Expenditure Method Exercised in Japan Applied to a Hypothetical Case of a Multi-Purpose Dam for Flood Mitigation, Irrigation and Hydropower Generation.

	Item	A: Flood mitigation	B: Irrigation	C: Hydropower generation	Total (¥ 10^9)
a	**Expected benefit**	50	35	150	235
b	**Alternative single-purpose cost (alternative justifiable-expenditure)**	40	60	100	200
c	**Smaller (a or b) (benefit limited by alternative cost)**	40	35	100	175
d	**Separable cost 1. exclusive facilities**	15	10	50	75
e	**Separable cost 2. common facility**	23	5	10	38
f	**Total separable cost (1+2)**	38	15	60	113
g	**Remaining alternative justifiable-expenditure c-f (remaining benefit)**	2	20	40	62
h	**In %**	3.2	32.3	64.5	100%
i	**Cost allocation of common facility**	1.6	16.15	32.25	K=50
j	**Total cost i+f**	39.6	31.15	92.25	163
k	**In %**	24.3	19.1	56.6	100%

cost – alternative justifiable-expenditure method suggests the following procedure. It is explained for user A but similar for B and C, too.

(a) Expected benefit: How much benefit can user A gets from this project P? The expected benefit that A can get from this project is the at-most amount that A can pay. Note that this is a gross benefit expressed in present value considering time horizon and discount rate. All costs referred in this procedure are present value.

(b) Alternative single-purpose cost (alternative justifiable-expenditure): How much does it cost for user A to obtain the same benefit by constructing a single-purpose dam or some other methods by A alone without joining this project P? This alternative cost is the upper limit that the user A can justify itself to pay for joining a project P.

(c) Benefit limited by alternative cost: User A can share the cost up to whichever smaller a or b.

(d) Separable cost 1. Cost of facilities for exclusive use: If project P includes some facilities that A uses exclusively, this cost should be shared by A. For example, flood control gates for A, water intake facility for irrigation for B and penstocks for hydropower generation for C.

(e) Separable cost 2. Additional cost for letting A join the project P: Other than cost for exclusive use, what is an additional cost for building the common facility by letting A to join the project as compared with the case where only users B and C build a dam? This additional cost should be paid by user A. For example, the additional dam height and size are necessary, and the work period becomes longer to include flood control capacity for A, and for B and C, irrigation water supply and hydropower generation, respectively.

(f) The total separable cost 1 and 2 should be paid by user A.

(g) Remaining cost share limit after paying the total separable cost: c-f. (remaining alternative justifiable-expenditure or remaining benefit)

(h) Remaining cost share limit in %.

(i) The cost share of the common structure of the dam K=¥50 10^9 by A should be allocated according to h.

(j) The total cost allocation: i+f

(k) Its allocation in %.

This is a traditional method used since the 1950s but is still of use now. Recent application example cases include the Shitara dam in Aichi Prefecture (Togashi, 2014). The basic stand of this cost allocation seems quite reasonable based on player's expected benefit bounded by alternative cost. But in practice, there may be many difficult problems such as the cases with no alternative solutions, urgency allowing no negotiation time, changes in targets, estimated costs and benefits, or even withdrawal over a long planning and implementation time, etc. (Hasebe, 2006). The methodology of estimating cost and benefits is also a difficult matter especially human lives, environmental quality, resilience, robustness, etc. are included. Therefore the role of coordination and negotiation in integration process cannot be replaced by such a tool but used as a guide of negotiation.

4.2.3 WMO efforts for Integrated Flood Management (IFM)

UN World Meteorological Organization (WMO) established in 1950 (preceded by an NGO International Meteorological Organization (IMO) established in 1873) has Hydrology and Water Resources Program (HWRP) established in 1961 (preceded by a Hydrological Commission (CHy) established in 1946 as an assembly of national hydrological services) which closely works with national meteorological services and hydrological sections responsible for rainfall measurement (Askew, 2008). The program interest is operational hydrology on measurement, forecasting, early warning and preparedness for floods and droughts that most of national hydrological services assume responsibility. This program started the Associated Program on Flood Management (APFM) in 2001 as a major contribution to member countries for flood management as follows.

Associated Program on Flood Management (APFM)

The APFM is a joint initiative of the WMO and GWP started in 2001 financially supported initially by Japan (2001–2014) and the Netherlands (2001–2010), and later by Switzerland, Italy, Germany, USAID, etc. It promoted the concept of Integrated Flood Management (IFM) as a new approach to flood management. The original form of APFM continued till 2017 (WMO, 2018) and now it is operating on a virtual basis focusing on "last miles" (APFM: website). The program published a series of supporting documents.

One of them is APFM Integrated Flood Management Concept Paper (WMO, 2009). This was the third edition of the one first published in 2003 that described the concept of integrated flood management.

To follow are the *author's interpretation* of the outline of the concept.

APFM definition of integrated flood management (IFM)

First the concept paper defined the IFM as follows:

> Integrated flood management (IFM) is a process promoting an integrated – rather than fragmented – approach to flood management. It integrates land and water resources development in a river basin, within the context of IWRM, and aims at maximizing the net benefits from the use of floodplains and minimizing loss of life from flooding.

This was an extension of IWRM definition to floods that included benefits of floods to be maximized and risks to be minimized. As mentioned in terminology in Section 4.1.1, however, APFM focus was floods as hazards and not resources. Therefore, APFM's IFM is equivalent to IFRM as obvious in their concept paper presented in the next section. Nevertheless, this textbook follows their use IFM as a terminology but it does mean IFRM.

APFM IFM concepts

The APFM concept paper (WMO, 2009) presented its new concepts in four steps: first describes the *traditional flood management* options, then points out the existing *challenges* and the *integrated flood management* to address the challenges, and finally *puts it to practice*. Here the traditional flood management options are not something wrong or inappropriate but simply the current exercises that need to be improved for better performance in efficiency and effectiveness by adding coordination to related components and sub-systems. They simply advocate that integrated management can make a total system perform better.

(1) Traditional flood management options

- Source control to reduce runoff (afforestation, infiltration, permeable pavements, artificial recharge)
- Storage to temporally adjust runoff (wetlands, detention basins, dam-reservoirs, retardation ponds)
- Increase channel capacity (channel deepening or widening, groin (groyne), spur dikes, riverbed works (groundsills, step works, check dams))
- Diversion canals (bypass channels)
- Urban sewerage system (storm water drainage, wastewater treatment)
- Separation of rivers from populations (land use control, dikes, floodproofing, zoning, house raising (pilotis))
- Emergency management during floods (flood warnings, emergency bank raise, emergency dike strengthening, floodproofing, evacuation)
- Flood recovery (recovery subsidies, debris clearance, counselling, compensation or insurance).

All those options are quite familiar to flood managers and are widely exercised. They all are fine and useful but there are many familiar limitations on how they are developed and operated. Such as:

- Ad-hoc and stand along (little attention to integrated river basin management, basin-wide use of multiple reservoir system, multiple purpose operation, etc.)

- Reactive rather than proactive (disaster response rather than risk reduction, water development after water saving becomes necessary)
- Emphasis on structural measures (lack of participatory approach, overconfidence of engineering potential)
- Mono-disciplinary (lack of joint work with such as water, energy, ecological, social, educational, cultural sectors)
- Failure to learn from the past (lack of precautionary approach)

IFM should overcome such limitations to address challenges that a society faces.

(2) The challenges of flood management

The challenges that the contemporary society faces and are difficult to be overcome without addressing the limitations mentioned earlier:

- *Rapid urbanization:* Urban population in the world is rapidly increasing by migration from rural area to urban areas, that is, 30% of the total 2.5 billion in 1950, 55% of 7.7 billion in 2019 and projected 68% of 9.7 billion in 2050 (UN, 2018, 2019). Most of population increase is expected to be in the coastal zone and flood-prone areas. Prevention of new risk creation is a real challenge in the world.
- *Climate variability and change:* Climate change is a new dimension to flood-related hazards. Intensity, volume, snow, geographical distribution, seasonal pattern, etc. all change with uncertainty and bring about great impacts. Adaptive approach with progress of climate and the balanced and the best mix of soft and hard measures and of control and adjustment to floods should be the strategy.
- *Ecosystem approach:* Floods have an important function to maintain ecosystem such as to provide fish spawning areas in floodplain, to help fish migration, and to flush debris, sediments and salt. They are particularly important in dry climate regions that experience seasonal flooding followed by a period of drought. Therefore, stopping high-frequency floods by dam reservoirs does damage ecosystem. Preservation of ecosystem is a prerequisite to sustainable development.
- *Impossibility of absolute safety from flooding:* However largescale infrastructure was constructed to prevent extremely low-frequency floods, even larger floods happen. Even for protecting against high-frequency floods, absolute safety is still an illusion. Some unexpected events or accidents cannot be avoided. They are the challenges to be prepared for.
- *Securing livelihoods for displaced people:* Floodplain provides excellent, technically easy opportunities of livelihood. Therefore, replacement of people from flood risk area to safer area needs to secure new livelihood for their living. Similarly, creating ponds, constructing levees or diversions that take up land use opportunities need the same.

(3) Integrated flood management: the concept

In order to address such challenges, IFM aims to realize *concepts described in IWRM and IFM definitions* in practice, namely, promoting coordinated management of involved components for integration. The principles for integration for IFM include the following:

- *Manage the water cycle as a whole:* Whole range of floods, small, medium and extreme; whole range of flood function, water use, hazards, sediment transport, groundwater recharge, etc.; whole range of flow networks, water supply, sewage, wastewater disposal and storm water drainage; whole range of management, freshwater, drought, coastal zone, etc. should be managed in total. They might have conflicting factors with each other and systematic coordination is indispensable.
- *Integrate land and water management:* Land use totally changes hydrology of a basin. Urbanization changes infiltration and retardation capacity of land and increases flood peaks and shortens flood concentration time, and the land use changes water quantity, quality and sediment yields as well as deposition. Thus land use and water management sectors should be closely coordinated through consultation and negotiation based on information exchanges on soil maps, land cover maps, hazard maps, etc. Development of smart cities (water-sensitive cities) is a result of integration by such sectoral coordination.
- *Manage risk and uncertainty:* Living and utilizing floodplain provides great opportunity but subject to risk and uncertainty. Stochastic nature of precipitation and climate variability is one thing and an even larger factor is societal changes such as by technology, culture, population, economy and politics. Any water management plan should be made resilient considering those risk and uncertainty in providing structural and nonstructural measures. Risk management process includes identification and assessment of risks, prevention and mitigation of risks to become a disaster and preparedness, response and recovery when a disaster occurs. Resilient society needs to develop capacity to overcome unexpected happenings.
- *Adopt a best mix of strategies:* Mixture of strategies and options of flood management should be sought for IFRM depending on natural and socio-economic conditions of the target area. Table 4.2 lists some examples of strategies and options. A best mix depends on particular conditions of localities that maximize the benefits from floodplain and minimize losses.
- *Ensure a participatory approach:* As Agenda 21 (UN, 1992) indicates the involvement of the public at all levels of decision-making and the role of women are important. In this approach, the process should be open, transparent, inclusive and interactive. However, a good

Table 4.2 Strategy and Options for Flood Management (WMO, 2009)

Strategy	Option
Reducing flooding (hazard control)	Dams and reservoirs Dikes, levees and flood embankments High flow diversions Catchment management Channel improvements
Reducing susceptibility to damage (exposure and vulnerability control)	Floodplain regulation Development and redevelopment policies Design and location of facilities Housing and building codes Floodproofing Flood forecasting and warning
Mitigating the impacts of flooding (preparedness and response control)	Information and education (hazard maps, disaster literacy and consciousness) Disaster preparedness (emergency response, evacuation and drills) Post-flood recovery Flood insurance
Preserving the natural resources of floodplains (preserving biodiversity and nature)	Floodplain zoning and regulation

combination of top-down and bottom-up is necessary. Too much bottom-up is a chaos. Responsible top-down from central administration has a critical role over local institutional efforts.

- *Adopt integrated hazard management approaches:* IFM should take a multi-hazard approach including not only landslides, slope failures, debris flows, windstorms, storm surges but also tsunamis, earthquakes, volcanic eruptions, etc. Disasters become large under unexpected conditions.

(4) Putting integrated flood management into practice

Finally, the concept paper describes the step for putting IFM into practice. The effective implementation of IFM and IWRM requires an enabling environment in terms of policy, legislation and information; clear institutional roles and functions; and management instruments for effective regulation, monitoring and enforcement. They all should be formed under interaction with local peculiarities. The following are examples of the necessary components to be considered to implement IFM.

- *Clear and objective policies supported with legislation and regulations:* Any IFM actions need political commitment with specific legislation and allocation of budget and other resources not only to the main leading agency but also other related institutions that make the institutional collaboration possible.

- *The need for a basin approach:* Development or changes in one place impacts other parts of the basin through water quantity, quality, sediments, nutrients, discharge patterns, geographical floodwater distribution, etc. Also, a basin development plan needs to be in accordance with a national development plan. Such spatial consideration and consistency are prerequisites to IFM.
- *Institutional structure through appropriate linkage:* IFM needs basin-wide institutional collaboration. But administrative boundaries do not coincide with a river basin where local governments should collaborate with each other. In case of transboundary rivers, it is not at all an easy matter. In fact, 90% of world population live in nations with transboundary rivers and 40% of world population live in transboundary river basins.
- *Community-based institutions:* For IFM stakeholders of community-based institutions are involved and necessary to be coordinated and cooperate across institutional boundaries to reach decisions and implementation at the basin level. But most local institutions do not have their decision-making process to facilitate basin-wide cooperation or community involvement in the "bottom up" approach. Such local institutions may need some training towards basin-wide IFM that includes a transparent participatory decision-making process.
- *Multidisciplinary approach:* IFM includes both benefits and losses of floods so that it necessarily involves multiple and often mutually conflicting socio-economic interests. Therefore, multidisciplinary and interdisciplinary approach is indispensable. In Chapter 6, transdisciplinary approach is advocated which emphasizes scientific decision-making under all disciplines and sectors working together till its final implementation.
- *Adaptive management:* As future flood risks are uncertain in climatic variability and socio-economic changes, IFM should necessarily be adaptive. It necessitates the solution should be robust and flexible to meet a certain range of unexpected natural and socio-economic conditions.
- *Information management and exchange:* IFM needs consensus building, for which all key data, information and experiences should be well shared in a transparent way. In order to follow such basic rule for collaboration, stakeholders may need training for IFM such as information sharing for upstream and downstream collaboration and emergency response collaboration.
- *Appropriate economic instruments:* A question of who pays and who benefits is an important societal understanding to be agreed upon. People living in the flood risk area have a primary responsibility to cover its cost. But such obligation and costs have to be open to and well understood by customers during real estate sales. Also, socio-political agreement of the responsibility of indirect beneficiaries who get benefit of flood plain activities have to be clearly understood and institutionalized through taxes, subsidies and insurance business.

REFERENCES

ADCP (2005) *A Primer: Integrated Flood Risk Management in Asia.* Asian Disaster Preparedness Center (AGCP). www.adpc.net/UDRM/primer/Volume%202/volume2.pdf (accessed 21 May 2021).

APFM: website www.floodmanagement.info/# (accessed 24 August 2020).

Askew, Arther (2008) Hydrology and Water Resources within WMO—the Birth of a Programme. *Bulletin,* 57(3). https://public.wmo.int/en/bulletin/hydrology-and-water-resources-within-wmo%E2%80%94-birth-programme (accessed 12 September 2020).

Bakker, Karen (2007) The "Commons" Versus the "Commodity": Alter-globalization, Anti-privatization and the Human Right to Water in the Global South. *Antipode,* 39(3), 430–455. https://doi.org/10.1111/j.1467-8330.2007.00534.x.

Barrow, Christopher J. (1998) River Basin Development Planning and Management: A Critical Review. *World Development,* 26(1), 171–186.

Butler, C. and N. Pidgeon (2011) From "Flood Defence" to "Flood Risk Management": Exploring Governance, Responsibility, and Blame. *Environment and Planning C Government and Policy,* 29(3), 533–547. http://doi.org/10.1068/c09181j.

European Commission: website on EU Water Framework Directive. https://ec.europa.eu/environment/water/water-framework/index_en.html (accessed 29 October 2021).

European Flood Directive (2007) Directive 2007/60/EC of the European Parliament and of the Council of 23 October 2007 on the Assessment and Management of Flood Risks. *Official Journal of the European Union,* L 288, 27–34. https://eur-lex.europa.eu/legal-content/EN/TXT/PDF/?uri=CELEX:32007L0060&from=EN (accessed 25 November 2020).

Fletcher, C., A. van Heelsum and C. Roggeband (2018) Water Privatization, Hegemony and Civil Society: What Motivates Individuals to Protest About Water Privatization? *Journal of Civil Society,* 14(3), 241–256. https://doi.org/10.1111/10.1080/17448689.2018.1496308.

Giger, Walter (2009) The Rhine Red, the Fish Dead—the 1986 Schweizerhalle Disaster, a Retrospect and Long-term Impact Assessment. *Environmental Science and Pollution Research,* 16(Suppl 1), S98–S111. http://doi.org/10.1007/s11356-009-0156-y.

Giordano, Meredith A. and Aaron T. Wolf (2003) Sharing Waters: Post-Rio International Water Management. *Natural Resources Forum,* 27, 163–171. https://onlinelibrary.wiley.com/doi/epdf/10.1111/1477-8947.00051 (accessed 1 February 2021).

Gleick, Peter H. (1998) The Human Right to Water. *Water Policy,* 1, 487–503. https://doi.org/10.1016/S1366-7017(99)00008-2.

GWP (2000) *Integrated Water Resources Management.* Technical Advisory Committee (TAC) Background Papers No. 4. www.gwp.org/globalassets/global/toolbox/publications/background-papers/04-integrated-water-resources-management-2000-english.pdf (accessed 20 November 2020).

GWP: website. www.gwp.org/ (accessed 17 May 2021).

Hasebe, Shunji (2006) *Water economy (8) Water Right and Dam.* Dam Japan, 740. June 2006. http://damnet.or.jp/cgi-bin/binranB/TPage.cgi?id=297 (accessed 26 January 2022).

ICWE (1992) *International Conference on Water and the Environment: Develop-ment Issue for the 21st Century*, 26–31 January 1992, Dublin, Ireland. www.irc-wash.org/sites/default/files/71-ICWE92-9739.pdf (accessed 27 January 2022).

IPCC (2012) *Managing the Risks of Extreme Events and Disasters to Advance Cli-mate Change Adaptation*. A Special Report of Working Groups I and II of the Intergovernmental Panel on Climate Change (SREX) [Field, C.B., V. Barros, T.F. Stocker, D. Qin, D.J. Dokken, K.L. Ebi, M.D. Mastrandrea, K.J. Mach, G.-K. Plattner, S.K. Allen, M. Tignor and P.M. Midgley (eds.)]. Cambridge University Press, Cambridge and New York, 582 pp. www.ipcc.ch/site/assets/uploads/2018/03/SREX_Full_Report-1.pdf (accessed 25 November 2020).

JMA Home Page on Terminology on Rivers, Floods, Heavy Rain Inundation and Ground Surface Phenomena. www.jma.go.jp/jma/kishou/know/yougo_hp/kasen.html (accessed 24 August 2020).

Lilienthal, David E. (1944) *TVA: Democracy on the March*. Harper and Brothers, New York.

Linsley, Ray K. and Joseph B. Franzini (1979) *Water-Resources Engineering*, 3rd ed. McGraw-Hill Book Company, New York.

Maass, Arthur, Maynard M. Hufschmidt, Robert Dorfman, Harold A. Thomas Jr., Stephen A. Marglin and Gordon M. Fair (1962) *Design of Water-Resource Systems*. Harvard University Press, Cambridge, MA.

Samuels, P., M. Morris, P. Sayers, J.D. Creutin, A. Kortenhaus, F. Klijn, E. Mossel-man, A. van Os and J. Schanze (2010) *A Framework for Integrated Flood Risk Management*. First Congress of the European Division of the IAHR at Edin-burgh, Scotland, May 2010. www.researchgate.net/publication/266880456_A_framework_for_integrated_flood_risk_management (accessed 26 August 2020).

Schanze, Jochen (2006) Flood Risk Management – A Basic Framework. In: J. Schanze, E. Zeman and J. Marsalek (eds.), *Flood Risk Management: Hazards, Vulnerability and Mitigation Measures*. Springer Nature, Switzerland, pp. 1–20.

Shiklomanov, Igor A. (1998) *World Water Resources, A New Appraisal and Assess-ment for the 21st Century. A Summary of the Monograph World Water Resources Prepared in the Framework of the International Hydrological Pro-gramme*. State Hydrological Institute, St Petersburg, Russia. https://unesdoc.unesco.org/ark:/48223/pf0000112671.

Showa 32 Governmental Ordinance 188 (1957) *Implementation Order of the Spe-cific Multi-purpose Dam Law*. https://elaws.e-gov.go.jp/document?lawid=332CO0000000188_20150801_000000000000000 (accessed 23 January 2022).

Togashi, Koichi (2014) *Water Resource Development Plan in Toyogawa and the Issues of Shitara Dam Project*. Gifu University, Faculty of Regional Sci-ence Research Report, 35, 101–126. http://repository.lib.gifu-u.ac.jp/bit-stream/20.500.12099/50289/1/reg_030035007.pdf (accessed 14 February 2022).

TVA Act (1933) *Tennessee Valley Authority Act of 1933*. The 73rd Congress of the USA. www.archives.gov/milestone-documents/tennessee-valley-authority-act (accessed 20 November 2020).

UN (1992) *Agenda 21*. United Nations Conference on Environment & Develop-ment, Rio de Janeiro, Brazil, 3–14 June 1992. https://sustainabledevelopment.un.org/outcomedocuments/agenda21 (accessed 26 November 2020).

UN (2018) *World Urbanization Prospects 2018*. UN DESA. https://population.un.org/wup/Publications/Files/WUP2018-Highlights.pdf (accessed 14 September 2020).

UN (2019) *World Population Prospects 2019*. UN DESA. https://population.un.org/wpp/Publications/Files/WPP2019_Highlights.pdf (accessed 14 September 2020).

UNESCO (2009) *IWRM at River Basin Level*. Co-edited by E.Z. Stakhiv and S. Ota. https://unesdoc.unesco.org/ark:/48223/pf0000186417 (accessed 21 November 2020).

UNGA (1992) *Report of the United Nations Conference on Environment and Development*. A/CONF.151/26 (Vol. I). Rio de Janeiro, 3–14 June 1992. www.un.org/en/development/desa/population/migration/generalassembly/docs/globalcompact/A_CONF.151_26_Vol.I_Declaration.pdf (accessed 27 January 2022).

UNISDR (2015) *Sendai Framework for Disaster Risk Reduction 2015–2030*. www.preventionweb.net/files/43291_sendaiframeworkfordrren.pdf (accessed 28 August 2020).

Vink, Karina (2014) *Vulnerable People and Flood Risk Management Policies*, PhD Dissertation. GRIPS, September.

WMO (2009) *Integrated Flood Management Concept Paper*. WMO-No. 1047. www.floodmanagement.info/publications/concept_paper_e.pdf (accessed 26 November 2020).

WMO (2018) *Draft Annual Report 2017–2018*. WMO APFM Report No. 43, August 2018. APFM-Draft-Annual-report-2017-2018.pdf (floodmanagement.info) (accessed 5 February 2021).

WWC (2004) *Analysis of the 3rd World Water Forum at Kyoto, Shiga and Osaka, Japan on 16–23 March 2003*. World Water Council and Secretariat of the 3rd World Water Forum. www.worldwatercouncil.org/en/kyoto-2003 (accessed 29 January 2022).

Chapter 5

Japanese experiences

In previous chapters, conceptual bases and policy evolution on disaster risk, and basic concepts on IWRM and IFRM were reviewed. Now based on such understanding, this chapter looks into concrete examples of IFRM in Japan. Japan is a country of all kinds of natural hazards located in typhoon-prone Asian monsoon region and the circum-Pacific orogenic zone with volcanic activities and earthquakes, resulting in experiences of many devastating disasters. Nevertheless, Japan has been managing her nation as much as possible going through incredible series of societal changes, wars, urbanization, industrialization, etc. one after another that may give some insights to many nations. Japan has been accumulating such experiences of disaster risk management since the dawn of history. This chapter reviews the late history of Japanese flood management, starting from hydro-environmental conditions, experiences before and after World War II and recent policy development on IFRM. In addition, tsunami experiences will also be touched upon. Those examples would show not only unique engineering technologies but also societal system that made such technologies and infrastructure operate and led to disaster culture of Japan. About the modern history of flood disasters and civil engineering in Japan, the work of late professor TAKAHASI Yutaka is outstanding such as (Takahasi, 1971, 1990). Views and interpretation of disasters expressed in this chapter especially Section 5.2 and 5.3 largely follow his with high respect and appreciation although the author bears full responsibility.

5.1 GENERAL HYDRO-ENVIRONMENTAL CONDITIONS

Hydrology and hydrological environment are a product of climate, land and human activities. Climate and land conditions determine human activities and, by return, human activities influence land and hydrological conditions. Japanese hydrology is no exception.

Climate: Japanese climate belongs to warm humid regions having distinct four seasons under major influences of Asian monsoon and typhoons. In the Köppen-Geiger classification, it belongs mostly to Cfa or Dfa/Dfb region, that

DOI: 10.1201/9781003275541-6

is, temperate, no dry season, hot summer, or continental, no dry season, hot/warm summer region (Beck et al., 2018). The average annual precipitation over Japan varies around 1700 mm/y. According to MLIT annual reports "Water Resources in Japan" since 1996 and "Current Water Resources in Japan" since 2015, based on about 1300 stations all over Japan, the average annual precipitation was 1718 mm/y over 1971–2000 (MLIT, 2005a), 1690 mm/y over 1976–2005 (MLIT, 2012b) and 1668 mm/y over 1986–2015 (MLIT, 2019a). The total amount is decreasing but the annual variation is increasing. This much precipitation is rather unique in the mid latitude around 30°–40°N where there are many deserts and semi-arid regions in the zone.

Japan is located at the northeastern edge of Asian Monsoon region having rainy seasons in April to May (Harusame), June to July (Baiu) and September to October (Shu-u), and snow in winter mostly from October to March, especially in the Sea of Japan side of the nation. Those four rainy seasons are brought by the move of monsoon fronts or by the alternation of the Pacific air mass and the Siberian air mass. When the air masses meet over Japanese Archipelago the stagnated rain front is formed and precipitates until the front goes up to the North in summer or down to the South in winter. When the Siberian air mass prevails, the world's heaviest snowfalls in the Sea of Japan side of the country. The highest record of accumulated snowpack was 11.82 m at Mt. Ibuki, Shiga Prefecture on 14 February 1945 (JMA: website on Historical Ranking of Extreme Records) and in the flat land 3.77 m in Takada, Niigata on 26 February 1945 (NAOJ, 2019).

In the Sea of Japan side of the Honshu island, the snowpack often reaches to the second floor of houses and remains for several months. People live underneath of snow. To keep normal life is difficult and needs repeatedly clearing snow off the roofs of wooden houses and roads. It is an extremely hard work for any residents especially for elderlies. Snow collected is dumped into water ditches which sometimes become overloaded and cannot melt the dumped snowpack to flow it away. To keep main roads clear for traffic, groundwater is often used to continuously melt snow on the road surface as its temperature is constant all year around but it consumes much groundwater and sometimes causes ground subsidence. Fighting against heavy snow is a difficult and costly exercise for any heavy snow communities.

Typhoons are another major source of precipitation, most often formed eastern off-shore of the Philippines from August to October. A typhoon is defined by Japan Meteorological Agency (JMA) as a tropical low pressure situated west of 180°E and north of the Equator with the maximum wind speed of more than 17.2 m/s (34 knots and wind power 8 in Beaufort scale) on average over 10 minutes (JMA: website on Typhoons). Typhoons occur, on an average, 25.6 a year over 1981–2010, among them 2.7 land on Japan and 11.4 come close (less than 300 km) to Japan. The largest number of typhoons landed on Japan was 10 in 2004 and the rest less than 6.

Typhoons bring strong winds and rainfalls. Both can result in serious damages, but rainfalls are usually more widely and seriously damaging than winds in Japan. Typhoon-associated rainfalls are not only from convective rain bands of typhoon itself but also by typhoon-stimulated frontal activities. The case of 2018 July Heavy Rain or West Japan Heavy Rain (both names given by JMA: website on list of JMA Named Natural Hazards) from 28 June to 8 July 2018 was this type brought by prolonged frontal activities stimulated by Typhoon 7, resulting dead and missing 245 persons (FDMA, 2019).

Thus, three major sources of precipitation, that is, Baiu, typhoon and snowfall form distinct four seasons and provide ample resources for forest, agriculture and all other vegetations in Japan. The rainfall intensity is especially high for several hours to several days' duration (nearer to the world records line) as seen in Kiguchi and Oki (2010). This fact is quite important to flash floods in Japan. As Japanese rivers are all small (the longest is the Shinano River 367 km) where flood concentration time is less than a few hours to at most a few days, the precipitation conditions are favorable to flash floods.

Land: Japanese archipelago is a long series of large and small islands with many high volcanic mountains, the highest peak being 3776 m of Mt. Fuji. The plate tectonics around Japan and the circum-Pacific zone create frequent earthquakes and sometimes tsunamis. There are the Median Tectonic Line from Kanto to Kyushu and the wide Fossa Magna whose western edge is sharp from Itoigawa to Shizuoka but the eastern edge is still under scientific debates ranging several tens to over a hundred km width. Other than those major tectonic lines, there are many other faults all over Japan.

Such geological conditions result in short and steep rivers in narrow basins producing huge amounts of sediments. Sediments are important material to fill the erosion along rivers, form alluvial fans, flood plains, delta around estuary and maintain coastlines. Sediments also supply nutrition to floodplain by floods to make agriculture possible and to the coasts to feed rich aquaculture. Under such geological and topographical conditions, 67% of Japanese territory is covered by forests in mountainous regions and 12.2% the cultivated land, of which 54.3% is rice paddy (NAOJ, 2019). Rice paddy provides ample retardation function for regional hydrology and creates rich biodiversity and beautiful landscape with distinct seasonality. Figure 5.1 shows typical Japanese cities in high population density located in narrow basins near flood-prone rivers, near erodible mountains, near the sea or in delta areas.

Human activities: As the Japanese main diet is rice, most Japanese have been living in lowland or flatland where rice paddy cultivation is possible. As lowland and flatland are floodplains, paddy fields and residential areas need to be protected from floods and debris flows while securing enough water for rice paddy during growth and domestic use during dry seasons. Human dwelling on the floodplain necessitated to fix the freely moving river courses by levees, diversion canals, retardation ponds, etc. and make water available when and where needed as much amount as necessary. Once levees were built, sediments settled within rivers and raised riverbeds which necessitated to make levees higher and higher repeatedly to prevent floodwater overflowing riparian lands. As a result, there are many rivers

Figure 5.1 (Top) Mt Fuji, Numazu city and the Kano River (Courtesy of MLIT) and (bottom) the Tsurumi River estuary (Courtesy of MLIT: website on the Tsurumi River). Japanese are typically living in high-density urban areas in narrow basins near flood-prone rivers in mountainous areas or in delta areas.

in Japan that flow above their adjacent riparian lands which are called "sky rivers" or "ceiling rivers". Some rivers became so high that there are cases such as in the Fuji River basin, Yamanashi Prefecture, the In River which flows over a traffic road (Takada tunnel in Ichikawamisato-chou), and the Gomyo intersection of rivers where three low local rivers (the Yoko, the Gomyo and the Nagasawa Rivers) cross below two high rivers (the Takizawa and the Tsubo Rivers) (Yamanashi Prefecture, 2005). There are many diversion canals, too, to let floodwater away from the populated area and safely release it to the sea.

Later, human activities expanded a great deal especially by urbanization and industrialization which created a countless number of modifications to hydrological processes such as dams, sabo works, weirs, intra- and inter-basin water transfers, underground rivers, water storages and infiltration facilities, land reclamation, deforestation, etc. They all change hydrological system and make it quite different from the original nature wherever human beings inhabit on the Earth. Hydrology is an integral phenomenon of all those components: climate, land and human activities.

Flood prone conditions of Japan

Japan is in a heavy rain region of the Earth with monsoons, typhoons and winter snowfalls. Japan is a country of volcanic islands with high mountains so that all rivers are steep and short, and flash floods discharge high flows and huge amounts of sediments. The floodplains occupy 10% of the total 377,900 km^2 Japanese territory, where 51% of the total 127 million population live and 75% of the total assets are concentrated (MLIT, 2003a). There are 577 km^2 zero-meter zone (area below the mean monthly high tide) in Tokyo, Ise and Osaka Bays where 4.04 million people are living (MLIT, 2005b). This is the area where business and industrial activities are concentrated and is subject to floods as well as storm surges.

The formation of the zero-meter zone was mainly by ground subsidence by consolidation of clay layer by over-pumping groundwater for industrial (especially cooling), agricultural and municipal water uses. This started in Tokyo as early as around 1900 and in Osaka around 1930 but the cause was not clearly known. It became certain as over-pumping of groundwater by land elevation survey conducted after the war indicated that during the period when all factories were destroyed and all industrial activities suspended by the air raids during the war, the land subsidence had stopped (Endo et al., 2001). In 1956, the industrial water act and in 1962, the law of groundwater pumping regulation for building use were enacted and the over-pumping of groundwater stopped that recovered the stability of land subsidence by around 1970. After that, groundwater level came back, and in some areas, the level came up too high and caused problems of underground structures lifting.

In the Niigata plain, pumping up of groundwater of water-soluble natural gas also made a significant contribution to ground subsidence in the 1950s and in 1959, regulation started and by the sixth regulation in 1973, the subsidence became under control (Niigata Prefecture, 2018).

In heavy snowfall areas, groundwater has been used for melting snow on roads which also considerably contributed ground subsidence. In Nagaoka City, Niigata Prefecture, the groundwater use for melting snow on roads started in 1963 and spread to other places in Niigata, Toyama and other prefectures (Fujinawa, 1988). Now it has been carefully regulated (Nagaoka City, 2017).

The depicted land profile in Figure 5.21 in Section 5.5.5 is a cross-section in Osaka prefecture between the Yodo River and the Yamato River, which is quite different from majority of rivers in the world with undulating topography in large flat lands such as the Thames in London, the Saine in Paris, etc. The formation of sky rivers is mentioned in the previous section. It clearly shows high flood risk in the Osaka Plain which is not exclusive there but shared by many large coastal cities in Japan.

5.2 A SHORT HISTORY OF FLOOD MANAGEMENT IN JAPAN

Japanese must have been living with floods at least since they started rice cultivation. Japanese history has been to a large extent a history of managing natural hazards including hydro-meteorological hazards, geological hazards, epidemics, etc. Among all, excess water hazards such as floods, landslides, debris flows and storm surges have been the most common and frequently visited.

5.2.1 Flood control works before the Meiji Restoration (1868)

The Meiji Restoration (1868) was the time when Japan restarted herself under the Emperor's reign ending the Shogunate reign. Japan reopened the country to the outer world in 1854 when Japan-US "Convention of Kanagawa" was agreed to open Shimoda and Hakodate harbors to the US ships that ended more than 200 years of national isolation started in 1639 when the Portuguese ships were banned and only China, Korea and the Holland ships were accepted. The Meiji Era (1868–1912) was the first period of catching up the western world. But before the Meiji Restoration, there had been many flood control works constructed with outstanding ideas. Some are still working now that deserve a special attention with respect.

Shingen Zutsumi (Shingen Bank)

The Fuji River has been called one of the "three steepest rivers in Japan" reflecting its importance for people's life and difficulty in navigation. About 3510 km² basin (excluding the Numa River basin that was connected by Hoshikawa diversion canal in 1974) is surrounded by 3000 m class mountains including 3776 m Mt. Fuji, the 128 km long river flowing down to the nation's deepest bay, Suruga Bay in the Pacific Ocean (MLIT, 2002). Besides, in the western part of the basin, the western ridge of Japan's major geo-tectonic fault, Fossa Magna (Itoigawa-Shizuoka line) is running which makes western river branches of the Kamanashi River and the Fuji River extremely sediment productive.

As a result, the Fuji River has been one of the most difficult rivers for flood control in Japan. Already in 825, according to a document in Asama Shrine, Ichinomiya, Kai country (now Fuefuki City, Yamanashi Prefecture), a large flood occurred in the Kamanashi and the Midai Rivers, for which a "Midai" (meaning an Imperial messenger) came down from Kyoto to Kai to dedicate the Sansha Shrine to the flood prevention God. This episode became the origin of the name of the Midai River (Kofu River Road Office: website on History of the Midai River). It has been said that gods from Ichinomiya Asama Shrine, Ninomiya Miwa Shrine and Sannomiya Tamamoro Shrine come and meet at the Sansha Shrine to pray for flood safety (Yamanashi Kanko: website on Sansha Shrine).

In 1519, 500 years ago, TAKEDA Nobutora moved his castle from Kawata, east of Kofu basin, to Tsutsujigasaki, north-central part of the basin in Kai country. This is considered as the foundation of Kofu City. As Kofu basin was frequently hit by devastating floods, in order to make the country Kai strong, it was absolutely necessary to control floods in both the Kamanashi and the Fuefuki Rivers, two main branches of the Fuji River. During the time of TAKEDA Shingen, a son of Nobutora, who ruled the country from 1541 to 1573, various flood control works were constructed and flood management methods were practiced, of which some are still working and exercised to date. The best known is the Shingen Bank, a flood management system at the confluence where the Midai River joins the Kamanashi River.

In 1542 when a major flood disaster occurred in the Midai River and the Kamanashi River, TAKEDA Shingen directed his vassal flood control works. The Midai River origins from the Donokoya Pass (1518 m) of the Southern Alps of Japan and flows down 18.8 km with lots of sediments from the western edge of the Fossa Magna, which form a wide alluvial fan and join nearly vertically to the Kamanashi River. In the alluvial fun area, the Midai River used to flow south to directly hit Ryuo District of the Kofu basin and often badly damaged rice paddies. To solve this problem, the core engineering ideas were as follows (points from A to G correspond to the sites in Figure 5.2):

(1) Fix the Midai River course, that had been freely moving over the alluvial fan, to one direction at the head of the fan by a series of spur dikes made of stones (A: Tsumiishidashi).

(2) Lead the river by stone made mounds (B, C: Shogigashira) to the open cut of the river terrace (D: Horikiri).

(3) Dissipate water energy by 16 big stones (E: 16-Koku-Seki).

(4) Join the main channel of the Kamanashi River so that the merged strong current hits a high cliff (F: Taka-Iwa), made of debris avalanche deposits (岩屑流gansetsuryu) from Mt. Yatsu volcanic collapse about 200,000 years ago (Kumai, 1994; Science Lab of the Daiichi High School, 2000).

(5) Control the reflected (bounced back) flows from the cliff by the strong well maintained Shingen Bank (G: Shingen Zutsumi) which was a series of Kasumite (open dikes) flexible to let high flows go out of the dike during floods by inundating paddy fields in the hinterland and after floods

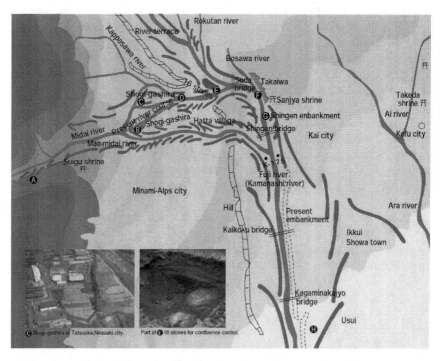

Figure 5.2 Flood Control works at the confluence of the Midai River to the Kamanashi River at the time of TAKEDA Shingen in the 16th century. The Midai River hits, after joining the Kamanashi River, nearly vertically the strong High Cliff made of debris avalanche deposits from Mt. Yatsu volcanic collapse. Note also, all dikes are Kasumite (open dikes). (Courtesy of San-nichi Press, Kofu, Kofu River Road Office: website on Traditional River Works)

to let return flows go back to the main river from open entrance downstream (H: open entrance for return flow at Ikkui and Usui). Note that in Figure 5.2, all dikes are Kasumite which absorb floodwater into back yard and return it back to the river after flood from other Kasumite downstream. Therefore, Kasumite were necessarily built in series.

(6) Maintain the Shingen Bank and other flood control system with the help of villagers whose real estate tax was waived as a compensation to their maintenance work including holding an annual shrine (Sansha-jinja) festival with a heavy miniature shrine (Omikoshi) carrying march (Omiyuki-san) over the bank shouting a unique cheering phrase "Sokodai-Sokodai" to compact it as shown in Figure 5.3.

Figure 5.3 The Three Shrines Festival (Omiyuki-san) to protect the Shingen Bank of the Kamanashi River, a branch of the Fuji River, held since 1557. They walk over the Shingen bank with steppes to compress it shouting "Sokodai-Sokkodai". (Top) The carriers wear red kimono and make up face in white which is considered not to embarrass goddesses. (Bottom) Along the riverbank, three groins are seen which are called "Seigyu" in Japanese meaning a sacred ox from its shape and origin from the Kamanashi River at the time of TAKEDA Shingen. (Photo: Courtesy of MOCHIZUKI Seiichi Collection, Yamanashi River Disaster Prevention Center)

The basic system of this safe merger of the Midai River with the mainstream of the Kamanashi River is still working and the concept of "using nature to control nature" (by keeping the original river flow as much as possible but taming it using nearby natural environment such as rock cliffs, open dikes, spur dikes, groins, etc. using local materials) named "Kohshuryu" (Kohshu method) flood control has been widely applied in Japan for river management. The other well-respected method has been "Kishu-ryu" which tries to fix the river flow by continuous levees, shortcuts etc. and keep floodwater away from the people while draining floodwater to the sea as soon as possible. Both have been used in many cases depending on local conditions.

The Tone River shift to the east

The Tone River (now the largest river basin in Japan, 16,840 km^2) used to flow into Edo (now Tokyo) Bay with many branches, of which two were the old Edo River and the old Sumida River. It now flows directly to the Pacific Ocean after the great civil engineering works called the Tone River shift to the east. Figure 5.4 illustrates how it developed spatially and chronologically.

The downstream of the Tone was unproductive muddy lowland and flood-prone area. Daimyo (feudal lord) TOKUGAWA Ieyasu came to Edo in 1590 (who eventually became Shogun in 1603) and requested his vassal INA Tadatsugu to improve the Tone River basin. Then the four generations of INA family (Tadatsugu, Tadamasa, Tadaharu and Tadakatsu) as Daikan or Gundai (the governor) of the Kanto District dedicated to the work. The Ainogawa, one of tributaries of the Tone River flowing down to the Edo Bay was closed in 1594 to protect paddy fields around and to improve the navigation to Edo, which made the Tone to flow to the east but still to Edo Bay. Later in 1621, since there was a need of navigation to transport northern Japanese products to Edo, further efforts of navigation started by excavating an inter-basin water transfer channel, the Akabori-gawa connecting the Tone to the Hitachi River, now the main route of the Tone River (Okuma, 1981). The work took a long time 1621–1654, but it made navigation from northern Japan to Edo through the Choshi outlet of the Tone possible. At the same time, by saving a large flood-prone area from flooding, it increased rice production a great deal. An estimate indicated the increase as much as 600,000 koku (90,000 t) increase (Takahasi and Sakou, 1963).

Thus, the original purpose of excavation of a channel for the Tone to flow east and to Choshi at the Pacific Ocean instead of southeast to Edo Bay was not for flood control but navigation. This was shown by Okuma (1981) from his thorough investigation of a huge number of historical documents on the

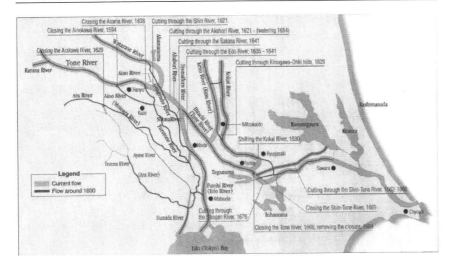

Figure 5.4 The Tone River shift to the east conducted in the 17th century. The original purpose was to promote navigation to transport rice and other goods to Edo. But it eventually became the main floodway to directly discharge floods to the Pacific Ocean instead of Edo Bay. (Courtesy of MLIT, Kanto Regional Development Bureau: Pamphlet "Tone River")

Tone River works. But nevertheless, the channel eventually widened to flow the major floods to east protecting paddy fields and saving Edo from floods. This contributed for Edo to become a center of economy in trade surrounded by large productive paddy fields. Although the major flood control works to protect Tokyo from floods should have waited till the Meiji Era, the large-scale inter-basin river work in the 17th century was one of the major civil engineering achievements then in the world (Archives Tone River Editorial Committee, 2001; JSCE, 2012).

5.2.2 Before World War II

At the time of the Meiji, Taisho and Showa Eras till right before World War II, modern technology was applied to flood control works responding to the need of socio-economic strengthening of the nation to catch up with the western powers as well as to support population increase, industrialization and militarization. Food control and water use development were part of fundamental practices to make it possible.

Invited foreign engineers

In the early Meiji Era, many foreign engineers were invited to lead the work for modernization of Japan with the advanced knowledge and technology.

Quite a few Dutch engineers helped flood control, river improvements, sabo works and harbor construction a great deal that included such as Cornelis Johannes van Doorn (stayed in Japan from 1872–1880) who helped to construct Azumi Sosui (canal), Nobiru harbor, etc., George Arnold Escher (1873–1878) for the Yodo River improvements etc. and Johannis de Rijke (1873–1903) for three Kiso rivers separation project and sabo works (Takahasi, 1990). As an influential teacher, Dr. William Smith Clark (July 1876–May 1877), President of Massachusetts Agricultural College, was probably best known who taught at Sapporo Agricultural College, Hokkaido. He was invited to Japan as he taught NIIJIMA Jo at Amherst College, Massachusetts as the first Japanese student and he introduced W. S. Clark to the Government of Japan. Dr. Clark accepted the invitation and stayed in Sapporo for 10 months. His successor was William Wheeler among whose students there were NITOBE Inazo, UCHIMURA Kanzo and HIROI Isami. Clark's famous legendary words are "Boys be ambitious! Be ambitious not for money or for selfish aggrandizement, not for that evanescent thing which men call fame. Be ambitious for the attainment of all that a man ought to be." which is written in a picture used to be hung in the clock tower of University of Hokkaido (now in the Hokkaido Old Prefectural Hall (Abe, 2014)).

To follow are some examples of basic flood control works implemented in the Meiji, the Taisho and the Showa Eras before World War II.

Shin-Yodo-gawa (the New Yodo River)

The Yodo River Basin (8240 km^2) is the 7th largest river basin in Japan having the Biwa Lake (669.3 km^2), the nation's largest lake, and major cities including Kyoto and Osaka. Osaka City, the second-largest city in Japan, is situated downstream of the Yodo River and was frequently hit by floods at the beginning of Meiji Era. Flood of 1885 was especially large and about a quarter of all houses in Osaka City were inundated and about 100 people died or missing (Yodo River Office: website on the history of the Yodo River). OKINO Tadao, the director of Osaka Civil Works Supervising Office submitted a proposal of the Yodo River improvement plan to then the Ministry of Interior that was accepted on the occasion of the 1st River Law of 1896. The main work was to open the New Yodo River (indicated by yellow in Figure 5.5(top) and Mark 3 in Figure 5.5(bottom)), the major 16 km diversion canal from Kema, in the north of Osaka City, to the Osaka Bay bypassing Osaka City from the west side. At Kema, a water gate with a lock (Kema Lock, Mark 2) was built to stop floods to flow down to the old Yodo River where the most populated part of Osaka-City was located. Including other related works, i.e., automated weir construction at the outlet of the Biwa Lake (Mark 4) to save both the lake and downstream cities from

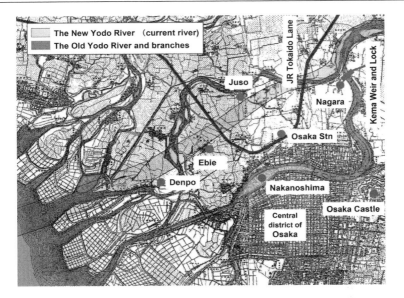

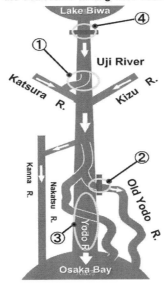

Figure 5.5 (Top) The New Yodo River opened and the Kema automated weir built in 1910 (Courtesy of MLIT, Yodo River Office: website on the History of the Yodo River), and (bottom) the Yodo River Basin Improvement Works Plan ①-④ by OKINO Tadao completed in 1910. ③ indicates that the Nakatsu River was divided into two parts by the new Yodo River. (Courtesy of MLIT, Yodo River Office: website on the Yodo River Reform)

floods, and replacement of the Uji River (Mark 1) to directly join to the Kizu River bypassing the Ogura Lake to mitigate its flooding, it took 14 years till 1910 to complete. Since then, Osaka City became much safer from Yodo River floods that helped its development a great deal as the second center of economy and business in Japan till now (Yodo River Office: website on the Yodo River Reform).

Okozu Bunsui (Okozu Diversion Canal)

Many diversion canals were built in the Meiji, the Taisho and the Showa Eras. The Okozu Diversion Canal (Okozu Bunsui) shown in Figure 5.6 (top) was one of the most representative one built from 1909 to 1931. Its purpose was to make the downstream Niigata Plain free from the Shinano River floods and to make many marshes into productive rice paddies. "Kata (gata)" in "Niigata" means marshes which did not suit to produce good rice and the produced was called "bird skipping rice (torimatagi)" meaning rice so bad that even birds do not eat and skip it over. The Shinano River is the longest river (367 km) in Japan whose upstream in Nagano Prefecture is called the Chikuma River (214 km). The diversion canal was built from the point about 50 km southwest upstream from the river mouth to Teradomari, the nearest coast about 10 km northwest. The canal started flowing in 1922 and after overcoming some accidents it was completed in 1927 but only a month later the foundation of the automated weir got fallen by land subsidence. AOYAMA Akira, Director of Niigata Civil Works Branch Office, Ministry of Interior and MIYAMOTO Takenosuke, Chief of Shinano River Repairment Office, took care of reconstruction and finally completed it in 1931. AOYAMA Akira left a memorial stone saying in Japanese and Esperanto: "Felicaj estas tiuj, kiuj vidas la Volon de Dio en Naturo. Por Homaro kai Patrujo. (Blessed are those who see the Will of God in Nature. For Humanity and the Fatherland)" This made muddy Niigata plain free from floods and one of the best quality rice (Koshihikari) producing country in Japan.

Figure 5.6(bottom) shows how large areas were saved from flooding by the Okozu diversion canal in case of the flood of September 1982. Later various counteractions by nature were also experienced such as serious erosion of Niigata Port and the coastline due to less sediment supply, riverbed rising of the original Shinano River by less sediment flashing, and time to time water shortage in some parts of downstream basin without floods. Now the original Shinano River is nearly a water supply and drainage waterway. On the other hand, the Okozu diversion canal became the main Shinano River channel and at its outlet, Teradomari area, new land or delta was created by ample sediment supply (Takahasi, 1990).

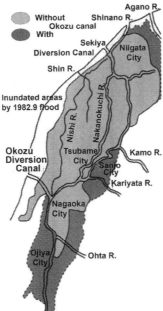

Figure 5.6 (Top) Okozu Diversion Canal, the wide left branch and the original Shinano River, the narrow right branch (Photo taken by Shinano River Office on 8 August 2018, Courtesy of MLIT). (Bottom) Its contributing area experienced in the case of flood in September 1982 (Courtesy of MLIT, Shinano River Office: website on Okozu Bunsui).

5.2.3 Chronological views of flood disasters after the Meiji Era

Regardless of such continuous flood management efforts, Japan encountered some floods every year somewhere. In some years they occur sporadically from one place to another and in some years covering wide areas. Many of them are floods exceeding the capacity of installed control facilities and the others are floods of different types that reflect the change of society.

Figure 5.7 indicates about 130 years of flood-related disaster losses. The black line is the number of people dead and missing and the red line is the economic losses in 2005 values. Some representative disasters were named which included the three greatest Showa typhoons marked with stars: Muroto the first, Makurazaki and Ise-Bay typhoons. It is noticeable that:

About human losses,

(1) In the center, the period between 1945 and 59 is named as the dark post-war 15 years when high death and missing rate, more than 1000 was experienced nearly every year.
(2) Before 1945, the end of World War II, the death tolls were rather high, time to time exceeding 1000 people.

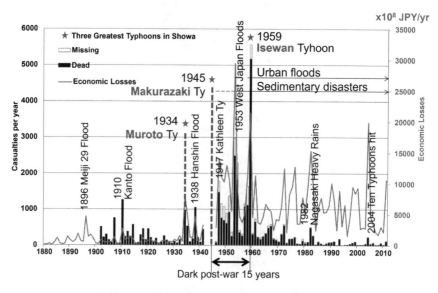

Figure 5.7 The chronology of the number of dead and missing and economic losses by water-related disasters (2005 value). Source: Water disasters statistics, 2011 (MLIT, 2014).

(3) After 1960, the level was getting smaller, always less than 1000, and recently seldom exceeding 200 but never gets zero.

(4) As flood control works of major rivers getting completed and urbanization proceeded, the affected areas of flood disasters shifted from along major rivers to small urban rivers and death tolls by sedimentary disasters such as landslides, debris flows and slope collapse in newly developed residential areas.

About economic losses,

(5) Compared with rather small economic losses before the war, they have been distinctly high after the war reflecting the economic growth. The mean annual losses have been about ¥5–600 billion (roughly $5–6 billion).

They reveal an important fact universal anywhere in the world, that is:

(6) Although death tolls decrease by various flood damage mitigation measures, economic losses do not. This corresponds to economic development with urbanization and industrialization of a nation and the shift of floods from extensive riverine floods to urban floods and sedimentary disasters.

Urbanization and industrialization increase economic vulnerability of society and flood damage potential (potential flood losses per unit flooded area). It is universal anywhere in the world, developed or developing. Increasing need of disaster risk management is therefore a fate of and an endless race with economic development.

The rest focuses on Japanese experiences of such never-ending efforts with societal changes. First, described are the period of post-war 15 years and then the recent flood management practices.

5.3 THE DARK POST-WAR 15 YEARS

It was a tragedy that the major disasters occurred following World War II, the weakest and the most damaged time of Japan and lasted as long as 15 years (1945–1959) referred as "the dark post-war 15 years". During this period, more than a thousand people were killed and more than 10% of national income was lost nearly every year. How and why did it happen?

Makurazaki Typhoon

It was on 17 September 1945, only a month (strictly 33 days) after the surrender of Japan to the allies of the United Nations Army, a super Typhoon Makurazaki landed on Makurazaki in the Satsuma Peninsula,

Southern Kyushu. The recorded air pressure in Makurazaki meteorological observation station was 916.1 hPa, the 2nd lowest air pressure observed in the history of typhoons landed on Japan, only next to 911.9 hPa of Typhoon Muroto that landed on the Muroto Cape of southeastern Shikoku Island on 21 September 1931. From Makurazaki, the typhoon crossed Kyushu Island to Shikoku Island, passed over Hiroshima City, went to the Sea of Japan and again landed on Tohoku and left to the Pacific Ocean (JMA: website on Makurazaki Typhoon; Hiroshima Prefecture: website on Makurazaki Typhoon).

In Hiroshima, it was only 42 days after the atomic bomb (A-bomb) was dropped on 6 August 1945 and totally destroyed the city as seen in Figure 5.8, having killed about 140,000 people by the end of 1945 (Hiroshima Prefecture: website on A-bomb Damages). This typhoon brought floods and debris flows, and it was reported that among 2473 people dead and 1283 missing in the entire Japan, 2012 were concentrated in Hiroshima Prefecture as many landslides occurred after deforestation during the war. Hiroshima casualties included 156 people in the A-bomb Hospital by debris flows of the Maruishi River, where patients were lingering on the brink of death, many doctors and nurses desperately helping them, and

Figure 5.8 Hiroshima after the A-bomb by US Army (Photo taken by US Army, Courtesy of Hiroshima Peace Memorial Museum). The Hon River (originally the Ota River) flows from the center bottom (North) to the right (Southwest) and a branch, the Motoyasu River from the center to the top (Southeast). The central bridge is the T-shaped Aioi Bridge which was the target of A-bomb drop.

the University of Kyoto team was investigating A-bomb impacts (Hiroshima for Peace: website). Many bridges survived from A-bomb in Hiroshima City were swept out and many houses which just started recovery were also swept out. On this day, September 17th Douglas McArthur started his occupational work at the GHQ in Daiichi Seimei Building in Tokyo.

Kathleen Typhoon

Another most serious flood and debris flow during this period was the Kathleen Typhoon that passed offshore of Shizuoka (the Enshu Sea) to the south of the Boso Peninsula, Chiba on 15 September 1947. It stimulated the stagnated autumn monsoon front and brought large amount of rainfall in the wide areas, especially the mountainous sides of Kanto and the Kitakami basin of Tohoku mainly on 14–15 September. Daily rainfall reached 519.7 mm at Chichibu, Saitama and 438.2 mm at Nikko, Tochigi on 15 September (JMA: website on Kathleen Typhoon). The total rainfall fell over large part of Kanto was over 300 mm and a number of deadly landslides and debris flows occurred. At Yattajima discharge observation station, Gunma, the flood peak of the Tone River was considered to have reached 17,000 m^3/s (other estimates say about 15,000 m^3/s), highest ever recorded in Japan (Matsuura, 2012).

By this flood, as shown in Figure 5.9(left), a large part of the Tone River basin was inundated. Figure 5.9(right) shows after the Tone River levee breached how floodwater advanced to Tokyo and reached Tokyo Bay. At the right bank of the Tone breached for about 340 m by overflow at Shinkawa-dori, Higashi Village (now Kazo City), Saitama, 4 km upstream of Kurihashi at 0:20 am of the 16th. The flooded flows flew down along the Old Tone River towards the Tokyo Bay joining various small rivers and inundating large areas. On the 15th, the Ara River left bank had breached for about 100 m at Kuge, Kumagai City, Saitama that joined to the flooded flows from the Tone River and eventually breached the Sakura Bank of the Ooba River on the 19th before dawn to flow into Tokyo. Inundating downtown Tokyo, the flooded flow finally stopped at the Shin River near the Tokyo Bay early 21st (Takahasi, 1971 CDMC, 2010). It took 5 days to flow down to the Bay. The route of the flooded water was the Old Tone River which was a branch of the Old Sumida River where the Tone River used to flow before its path was shifted to the east to the Pacific Ocean in the 17th century.

The number of people died and missing were 1077 and 853, respectively, of which the majority were lost along the Watarase River (a left branch of the Tone River) in Gunma and Tochigi Prefectures especially by landslides and debris flows in the area of Ashio hillsides and the Akagi mountain and about 300 m bank breach in the lower reaches of the Watarase

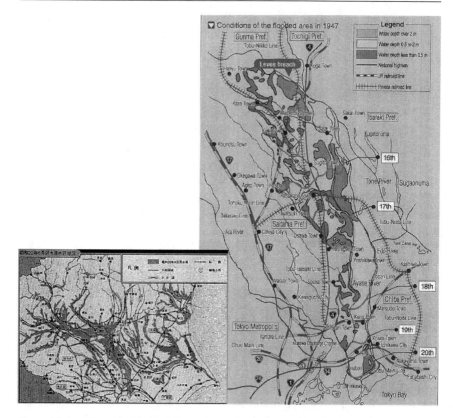

Figure 5.9 Inundation by Kathleen Typhoon (left) flood inundation areas (Courtesy of MLIT, Kanto Regional Development Bureau: website on Kathleen Typhoon Special) and (right) the advancement of flood inundation front after the dike breach of the Tone River at Higashi Village, Saitama on 16 September 1947. (Courtesy of MLIT, Kanto Regional Development Bureau: Pamphlet on Tone River)

River before joining to the main Tone River. Both landslides and bank breaches were considered as results of devastated forests and mountains by smoke pollution with sulfurous acid gas from the refining process of the Ashio Copper Mine in the head water area of the Watarase River that was founded in 1610 and became highly productive after the Meiji Era (CDMC, 2010).

The inundated area below Higashi Village, the right bank of the main Tone River was 440 km², the number of houses collapsed or half-collapsed was 31,000 and inundated was over 300,000 (CDMC, 2010). It was the severest flood that Tokyo has ever experienced in her history. This disaster initiated the flood control plans including dam construction upstream of the Tone and the Kitakami Rivers.

Ise-Bay Typhoon

The deadliest typhoon disaster during the post-war 15 years was at the time of the Ise-Bay Typhoon (Typhoon 15) with winds, storm surge and floods that gave severe damages all over Japan except Kyushu, especially the Ise-Bay area including Nagoya City on 26 September 1959. It landed on the west of the Shiono Cape of Wakayama Prefecture, the Kii Peninsula with 929.5 hPa (3rd lowest in the recorded history in Japan at that time). The tidal height was TP3.89 m at Nagoya harbor (TP (Tokyo Peil) is the mean sea level in the Tokyo Bay), the highest ever recorded in its history (CDMC, 2008). The most important characteristic of this typhoon was the number of casualties, that is, the dead and missing were 5098 (4697 and 401, respectively), the largest ever after the Meiji Era. The damaged houses were about 834,000 and the inundated area was 3636 km^2 (NAOJ, 2019).

The southern part of Nagoya City was badly damaged by overlapping effects of storm surge, river flooding and inland floods where damages were amplified by land subsidence of the reclaimed lowland mainly due to over-pumping of groundwater under rapid industrial development. Still worse, as the Atsuta (Nagoya) harbor had been historically famous on shipment of Kiso cypress, there were then many lauan timbers imported from overseas kept in stockyards for plywood factories from where timbers flew out and hit houses and people that took many lives (Takahasi, 1990).

An Extraordinary Session of the Diet was held in October 1959 and passed 27 acts for special responses to disasters, followed by the "Basic Law of Counter Measures for Disasters (BLCMD)" in 1961, and the Ministry of Education (then) started the Grand-in-Aid program for disaster science comprehensive research in 1960 (CDMC, 2008; Takagi, 1997).

Since 1960, there has been no disaster sacrificing more than four digits (1000) people until the Hanshin-Awaji Earthquake on 17 January 1995 and the Great East Japan Earthquake and Tsunami on 11 March 2011. The human losses due to a flood-related disaster are seldom more than 200 after the 1982 Nagasaki Heavy Rain but annual losses never become zero (e-Stat: website on Water-related Disaster Damages).

Other flood-related disasters during the dark post-war 15 years

There were many other flood-related disasters one after another during this period which were made even more tragic by four major earthquakes right before and after the end of the war (Takahasi, 1990). They are listed as follows: Note that the dates of hazards and the basic disaster loss data are after NAOJ (2019).

7 December 1944: The Southeastern Sea Earthquake of M7.9. The epicenter was offshore of Kumano Sea of the Kii Peninsula. Tsunami

hit many places especially the Mie coast only 10 minutes after the quake. 1223 persons died or missing (some say 589 by Tsunami in Mie), 17,599 houses totally collapsed and 3129 were washed away (CDMC, 2007).

13 January 1945: The Mikawa Earthquake of M6.8, an inland earthquake with the epicenter directly below Mikawa Bay, Aichi Prefecture. Some say it was an aftershock of the Southeastern Sea Earthquake. Dead and missing were 2306 persons centered in south of the Okazaki plain and the Sangane mountains.

Great damages of both disasters were kept secret from the people as the military government had to prepare for the homeland battle. But they heavily damaged military industry of Nagoya and Tokai areas which were considered negatively influenced the continuation of the war. (Takahasi, 1990; CDMC, 2007)

17–18 September 1945: Typhoon Makurazaki, introduced earlier in this section.

9–13 October 1945: Typhoon Agune landed on Kagoshima, Kyushu and the dead or missing were 451 people. The rice harvest in 1945 was poorest since 1902, the year of cool summer and typhoons (e-Stat: website on Crop Production; Fujibe, 2018).

21 December 1946: Southern Sea Earthquake of M8.0. The epicenter was Kumano Nada offshore Kii Peninsula. Tsunami hit seriously in Kochi, Wakayama and Tokushima prefectures. 1330 people died or missing, 11,591 houses totally collapsed, 1451 washed away and 2598 burnt out.

14–15 September 1947: Typhoon Kathleen, introduced earlier in this section.

28 June 1948: Fukui Earthquake, M7.1, the epicenter was in Fukui City. 3769 persons were killed, 36,184 houses totally collapsed and 3851 burnt out.

15–17 September 1948: Typhoon Ion passed a similar path of Typhoon Kathleen and damaged Kanto and Tohoku again. Dead and missing were 512 and 326 persons and 5889 houses were totally collapsed. Central part of Ichinoseki city, Iwate Prefecture was inundated in two consecutive years. (JMA: website on Typhoon Ion)

20–23 June 1949: Typhoon Della landed on Kagoshima and passed over Kyushu to north. Strong winds sank many fish boats in the Uwa Sea and a passenger boat Aoba-maru in the Suoh Sea near Hime Island of Oita Prefecture. 252 were dead, 216 missing. (JMA: website on Typhoon Della).

31 August to 1 September 1949: Typhoon Kitty landed on west of Odawara, Kanagawa Prefecture and left for Sado Island, Niigata

Prefecture which brought high tide in the Tokyo Bay, highest since 1917 and sunk 26 ships among which 90 was in the berth in Yokohama harbor. The salinization damage reached deep in the Sagami plain of Kanagawa Prefecture. Dead and missing were 160. (JMA: website in Typhoon Kitty; Kotobank: website on Typhoon Kitty)

2–4 September 1950: Typhoon Jane landed on Shikoku Island, passed Awaji Island, near Osaka and left to Wakasa Bay of Fukui Prefecture. The high tide in Osaka Bay caused the largest damage since 1934 Typhoon Muroto. It was recognized that the western low land area of Osaka city was 100–150 cm subsided. 508 people died or missing in Japan including 256 persons died or missing in Osaka Prefecture. A total of 19,131 houses totally collapsed. The experiences of high tide by Typhoons Kitty and Jane initiated the regulation of groundwater over-pumping and the promotion of industrial water supply. (JMA: website on Typhoon Jane; MLIT: website on Typhoon Jane; Ministry of Environment, 1971)

13–15 October 1951: Typhoon Ruth landed on Agune, Kyushu passing Yonago, Tottori Prefecture and Niigata, left to Ishinomaki, Miyagi Prefecture. 943 people died or missing mainly by slope failure around the area of withered granite (Masa and Sirasu) regions of Kagoshima and Yamaguchi Prefectures. In the latter, a whole village was destroyed and for the first time the National Police Reserve joined help operation. (JMA: website on Typhoon Ruth; Yamaguchi Prefecture: website on Typhoon Ruth).

25–29 June 1953: West Japan Floods. Torrential rains caused by a monsoon front resulted in unprecedented floods in the Chikugo and Shirakawa rivers in Kumamoto Prefecture, debris flows in Moji, Fukuoka Prefecture and inundation of the Kanmon Tunnel between Honshu and Kyushu Islands (Takahasi, 1971). Over 1000 people died or missing.

16–24 July 1953: Southern Kii Peninsula Torrential rains caused serious floods especially in Wakayama Prefecture and the dead or missing became over 1000. Again in September, Typhoon 13th hit Tokai region. The Cabinet office established the mountain and flood control coordinating council, issued the mountain and flood control plan and promoted the multi-purpose dams concept (Takahasi, 1990).

25–27 September 1954: Toyamaru Typhoon. The typhoon regained power in the Sea of Japan and hit Hokkaido with very high speed. 1761 persons died or missing of which 1139 were by sunken Toyamaru Ferry boat that was transporting trains from Hakodate, Hokkaodo to Aomori, Honshu. 5581 ships including 3 train ferries in the Hakodate harbor were damaged and Tokachimaru, another train ferry mooring outside of the harbor sunk. Due to strong wind, many trees fell down in the Ishikari river basin and 80% of Iwanai-town burnt out by fire. The number of deaths was the third biggest in the world history of disasters in the sea after the Titanic and the Sultana. After the marine court judgment, ferry transportation during storm warning became more reluctant and accelerated the construction plan of Seikan tunnel between Honshu and Hokkaido (Takahasi, 1990).

25–28 July 1957: Isahaya Torrential rains caused a serious flood disaster that destroyed newly developed residential areas in Isahaya, Nagasaki Prefecture. The dead or missing were 992 and it is considered as the first urban flood in Japan (Takahasi, 1971).

26–28 September 1958: Typhoon Kanogawa caused serious floods in Shizuoka Prefecture and Southern Kanto, and urban floods in Tokyo and Yokohama where rice paddies were reclaimed into residential areas or other open spaces were newly developed to residential areas. The dead and missing were 1269 persons. Ever since, urban floods spread into all over Japan as if they run after urbanization (Takahasi, 1990).

26–27 September 1959: Typhoon Ise-Bay, introduced earlier in this section.

Reasons of the dark post-war 15 years

For reasons why so many flood-related disasters occurred one after another during the post-war 15 years, there are several factors considered as follows:

1. It was unfortunately a relatively wet period in climatic variation. As seen in Figure 5.10, the mean annual rainfall over 46 capital cities in Japan was indeed continuously high in this period as compared with other periods although the annual values considerably vary. From the figure, it is also identifiable that: i) the national average decreases with the rate of 50 mm/100 years, ii) the recent fluctuation width is considerably increasing indicating notable drought years such as the 1978–1979 Fukuoka drought, resulting in 287 days' water supply rationing, and 1994 Heisei drought (sometimes called Archipelago drought) with many rationing in various cities including 295 days in Fukuoka again (MLIT, 2006).

2. During the war, little budget, if not none, was allocated for the maintenance and the management works of rivers (See Figure 5.11). They made rivers and the institutional arrangements weak against floods. Still worse, it was the time when the construction of continuous levees (normally from downstream to upstream) was

completed in many rivers and flood discharge downstream increased with reduced overflows and retardation upstream. It was the time more care was necessary to accommodate the new conditions. For instance, in the Tone River, a major reform started in 1900 after the major flood disaster in 1896 which was repeatedly revised later mainly with higher levees after consecutive flood disasters in 1910 and 1935. The 1940s were about the time such efforts were getting completed (Takahasi, 1971).

3. Meteorological observations and forecasting technology were yet primitive and, still worse, the information transmission and dissemination were so bad that little warning and preparedness were possible. The observed low pressure of Typhoon Makurazaki in 1945 was not transmitted even from Makurazaki to Meteorological Agency in Tokyo as the telephone line was all destroyed yet (Yanagida, 1975).

4. Another serious factor was the people's feeling about the value of human life. During the war, three million Japanese died and in such a tragic and inhuman situation, people were less sensitive to another tragedy.

Conversely, the reasons why no 4 digits losses occurred after 1960 were that all those situations got improved. The climatically wet period was over, more budgets were allocated for river management, hydro-meteorological observations, forecasting and information dissemination were improved, and society became more sensitive to human lives.

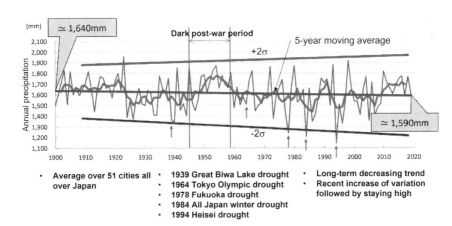

Figure 5.10 Chronological changes of annual precipitation averaged over 46 capital cities of prefectures based on Current Water Resources in Japan, 2020 (MLIT, 2021a).

5.4 CHALLENGES OF FLOOD MANAGEMENT AFTER 1960

Flood risk management that has been practiced since the latter part of the 20th century in Japan is a result of long-term modernization efforts of water and related land management since the Meiji Era. It is not only for flood risk management but also for more general resource and environmental management. The chronological view of Japanese water and land-related laws and policies illustrates the evolution of main interests and necessity of the nation.

River laws

There have been three river laws enacted since the Meiji Era. They respond to societal needs of the time.

The first river law, often called as the Old River Law, was enacted in 1896 (Meiji 29), which declared that

(1) river zone and river water are the public property,
(2) local government assumes the responsibility of river management except transboundary issues and technically difficult problems which would be taken up by the minister-in-charge of the national government,
(3) new users need to get license from local government but what farmers have been traditionally using for irrigation is given a vested right, and
(4) local governments have responsibility to appropriate budgets for their work but the national government shares some or all for the exceptional cases and subsidizes major works that exceed local governments' capacity.

At the beginning of the Meiji Era, the national government's priority was navigation (low flow works) but it soon shifted to flood control by local governments' requests. The law was immediately implemented for flood control works in the Yodo and the Chikugo Rivers, and then the Kiso, the Tone, the Shou and the Kuzuryu Rivers by 1900 (Yamamoto and Matsuura, 1996).

The second river law, often called as the New River Law, was enacted in 1964 (Showa 39), that was enacted 68 years after the old river law and 19 years after the War ended, during which big socio-economic change occurred and revisions were necessitated.

(1) One of the fundamental changes after the War was an introduction of decentralized governance system. Namely, in 1947, the new Local

Autonomy Law was enacted and the prefectural governor which was a position appointed by the central government became a position directly elected by the people in the prefecture. Accordingly, the prefecture became to share a distinct right in administration independent from the central government. The new river law, reflecting this change, the national government's responsibility was more clearly set. Namely, all rivers in Japan were classified into three categories: Class A rivers that are especially important for nation's conservation or economy are designated and managed by the Minister-in-charge of the national government, Class B rivers that are important are designated and managed by Prefectural Governors, and all other rivers out of the River Law regulation are managed by CTV Governors. There have been 109 class A rivers which are more than 1000 km^2 multi-prefectural rivers. The river manager's responsibility includes both flood control and licensing for water use. The latter meant to take the prefectural governors' right away and give it to the central government.

(2) The other big change of society was the ever-increasing needs of water use reflecting the economic development of the nation. The New River Law then took the principle of the integrated management of flood and water use. There were conflicts of water withdrawal and storage for hydro-electric power, municipal water supply including domestic, business/public and industrial water uses, and flood control. This created conflicts between upstream and downstream and among competing water users. Especially for storage dams, the licensing procedure was clearly set with conditions such as no negative impacts to original users and pre-registration of dam operation rule not to give negative impacts to downstream.

(3) The third point was the principle of basin-wide comprehensive management from the upstream to downstream for both flood control and water use. It was directed in the River Law Implementation Act of 1965 that the "basic high flow" at the time of the design flood should be identified in the "Basic Plan for Implementation of Construction Works" by the river manager (the minister-in-charge for Class A rivers and prefectural governors for Class B rivers). This basic high flow is the natural discharge by the design flood, which should be allocated to dams or retardation ponds or the river for flood control to be identified in the "river works implementation plan". The allocated flood flow at each river section is called the "planned high flow" for which levees and channel improvement works are constructed (Kajiwara, 2021). Designation of "basic high flow" in all river basins necessitated the use of hydrological statistics for balanced designation of safety level in different rivers according to their relative importance in Japan.

An example of the serious needs of water resources development would be highlighted by the case of Tokyo at the time of the Tokyo Olympic Games in 1964. It was during a long-lasting drought period in the Tokyo area. Low precipitation in the Kanto region started in 1960 and continued more than three years. All the reservoirs eventually became nearly exhausted and 20% water supply rationing in Tokyo started in October 1961 that was strengthened to 35% in July 1964 followed by 45% in August and temporarily reached 50%. It was called "Tokyo Desert" and the rationing lasted as long as 1259 days (3 years and a half) till 1 October 1964. This situation was relieved finally by a blessing of long-awaited rainfall on 20 August 1964 in the Kanto area followed by emergency opening of the Asaka Canal on 25 August to intake the Ara River water to Tokyo (from the Asaka Water Treatment Station to the East Murayama Water Treatment Station to supply water) that saved the Tokyo Olympic Game started on 10 October 1964 (Takasaki, 1996; Wakabayashi, 2015).

As a more fundamental solution, an inter-basin water transfer canal, the 14.5-km-long Musashi Canal was constructed to transfer water from the Tone River at Gyoda City to the Ara River at Konosu City. The construction started in January 1964. From the Ara River, the Asaka Canal was constructed to connect the Akigase Intake Weir to the Asaka Water Treatment Station. It was completed in March 1967 (JWA: website on the Tone River Canal).

The third river law, often called as 97 Amended River Law was enacted in 1997 (Heisei 9). The new foci were environment and public participation. It clearly says as follows (Omachi, 1998):

> Article 1. Purpose of this Law is to . . . maintain the normal functions of the river water by maintaining and conserving the fluvial environment.
>
> Article 16–2–4 . . . river administrators shall take necessary measures, such as public hearings etc., to reflect the opinions of the people concerned whenever necessary.

It requests each basin to prepare, replacing the no time-bounded Basic Plan for Implementation of Construction Works of the 2nd River Law, the "River Management Basic Policy" to make a long-term vision clear on flood control, water use and environment, and its implementation procedure to be completed within 20–30 years, the "River Management Implementation Plan". Both the policy and the plan should be determined in agreement with the heads of local governments and the River Management Implementation Plan needs to have the agreement of local residents such as through public hearings (Kinki Regional Development Bureau:

website on River Management Basic Policy and River Management Implementation Plan).

As indicated in Table 5.1, following the enactment of the first River Law, Sabo (sediment control) Law and Forest Law were enacted next year in 1897 (Meiji 30). Those three laws are called *Flood Control Triple Laws*. It indicates that in order to tame rivers it is necessary to control sediments from mountain slopes and conserve forests to smooth out runoff generation. This judicial scheme well reflects the volcanic mountainous conditions of Japanese rivers.

Following river laws, the important policies were adopted by the government in relation to integrated flood risk management. In the recent half-century, they were the Comprehensive Flood Control Measures in 1977 and the Effective Flood Management Including Basin Resistance to Floods in 2000 (see Table 5.2). These policies are virtually the Japanese declaration of integrated flood risk management (IFRM) as part of integrated water resources management (IWRM). More detailed judicial arrangements for IFRM will be discussed in Section 5.5.1 with Table 5.2.

Table 5.1 River Laws, and Sabo and Forest Laws, the Bases of River Basin Management in Japan.

Year	Law	Emphases
1896	1st River Law (Meiji 29 Law No. 71, 1896)	River is public property. Securing farmers' traditional use as the vested water right. Principal responsibility of flood control resides in Prefectural Governors. Inter-prefectural flood control is national responsibility.
1897	Sabo Law (Meiji 30 Law No. 29, 1897)	To regulate the use of land and provide materials for sediment control.
	Forest Law (Meiji 30 Law No. 46, 1897)	To conserve forests to prevent sedimentary disasters, protecting water sources etc.
1964	2nd River Law (Showa 39 Law No. 167, 1967)	Class A rivers designated and managed by the Minister and class B rivers by Prefectural Governors. Both for flood control and water use for development. Comprehensive basin-wide management. Basic Plan for Implementation of Construction Works specifies the design flood.
1997	3rd River Law (Amendment to 2nd River Law) (Heisei 9 Law No. 69, 1997)	Environmental conservation. Local governments and citizen participation. River Management Basic Policy and River Management Implementation Plan.

References are given in parentheses ().

Before getting into the explanation of IFRM in Section 5.5, it is necessary to see more about the consequences of flood control efforts after the war.

Flood losses and flood control investment

Figure 5.11 shows the 120 years chronology of Japanese flood-related disaster losses (black line) and flood-related disaster management investment (pink line) in proportion to national income expressed in year 2000 value.

From this diagram, it is observable that there are two peaks of high economic losses in the 1890s and the mid-1940s to 1950s and that the data are missing from 1941 to 45 due to World War II. The economic losses have been in the range of 0.1–10% of national income while the flood management investments 0.2–0.9%. The damages were much bigger than the investment during the Meiji Era to the mid-Taisho Era (till the early 1920s). In between the early 1920s to early 1930s, they are about the same, and mid-1930s to early 1960s, that is, during the influence of World War II,

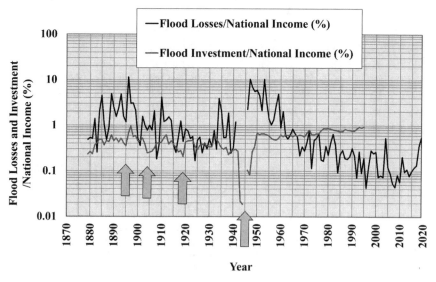

Figure 5.11 Annual flood-related economic losses (black) and flood-related disaster management investment (red) relative to national income in Japan

NOTE:

1) Economic damage = private property damage + public property damage + public service damage

2) National income indicates gross national income (GNI) which is the total amount of income earned by a nation's population or business in a year which is different from the gross national product (GNP) that is goods and services produced in a nation in a year.

3) Source: MLIT (2021b) Water disasters statistics, 2019.

again the losses were much larger than the investment. But the rest after the mid-1960s, the losses decreased remarkably reflecting the improvement of flood management infrastructure and the rapid economic growth of the nation but the investment has been kept high approaching near to 1% of the national income.

This indicates how Japan has been suffering from and struggling with flood-related disasters. Ten percent of national income is a huge loss in national economy and a major hold back to development. It also indicates that even if the loss reaches nearly 10% of national income, the investment to prevent or recover from the devastation cannot be more than several tenths of percentage up to nearly 1% of national income. National budget cannot be so large and solely allocated to disaster management but to many other targets.

Also, it indicates that flood management is an endless process and that the recent flood management investment much exceeds the average annual losses. This is because the flood losses were kept low by large continuous investment to flood management and if it was cut, flood losses would increase again. It is similar to the self-defense budget which is always much more than actual losses in order to keep avoiding wars. The arrows in the figure indicate the times when the investment dropped from the vicinity that correspond to the time when wars occur, namely, the war against China in 1894–1895, against Russia in 1904–1905, World War I in 1914–1918 and World War II in 1941–1945. During the war or the recovery from the war, most budget was allocated to its execution and, naturally, little was left for flood management. This by return indicates that if one nation commits any war (including civil wars), she cannot afford to invest for flood management. She may be said even unqualified to seriously consider disaster risk reduction from natural hazards.

Why economic losses do not decrease despite that human losses decrease?

Figure 5.11 indicates that the human losses can be reduced but not economic losses. Human casualties can be reduced by development of infrastructure, its maintenance and preparedness, especially early-warning and evacuation although not absolutely to zero. This is an obvious result of continuous investment of budget and human resources. But the economic losses do not decrease in nominal value although its proportion to the national income decreases reflecting an increase of national income. Why is it so? What is the reason to keep economic losses unreduced? Figure 5.12, a result of MLIT investigation indicates the reason. It shows that the total inundated area in the whole Japan decreased, from 1970 to 2001, from about 2000 km^2 to 400 km^2, one-fifth and the inundated area

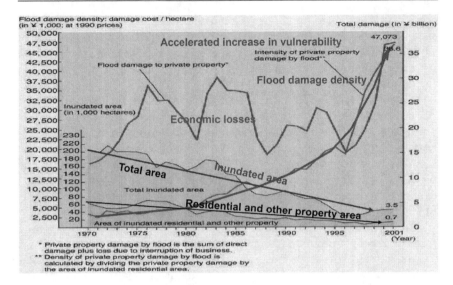

Figure 5.12 Accelerated increase of flood vulnerability in terms of flood damage potential from 1970 to 2001 in the whole Japan. Inundation area decreases but not economic losses because of increase of vulnerability. (Courtesy of MLIT)

of residents and properties from 600 km² to 100 km², one-sixth. But as the red broken line indicates, the economic losses, although fluctuating, did not decrease. It means that flood damage potential, that is, property or activity damage per unit inundated area increased by 5–6 times. In other words, flood vulnerability to inundation increased so much during these about 30 years.

Such change is a result of economic development and associated urbanization and industrialization. As flood control works progress with major infrastructure such as diversion canals, levees, dam reservoirs, etc., the main flood risk area shifts from extensive rural agricultural area to human and value intensive urbanized areas which results in increase of flood damage potential. This is a universal phenomenon occurring everywhere in the world and continues even after economic development as long as value-intensive properties and activities keep increasing.

The flood damage potential per unit inundation area increases with the increase of property values such as electronical facilities, intelligent houses, quality furniture, etc., and sophisticated business activities. Many commercial goods and business facilities are becoming more valuable and expensive. Various services such as transportation and communication become sophisticated and highly susceptible to an unexpected disruption. Production processes of various goods and services are becoming interdependent with each other forming supply chains and value chains in national and international wide. If one component stops in production process, all mutually dependent business activities in a market are affected, may be stopped till an alternative way is made available or adjusted to without it. Thus, business continuity planning (BCP) is imperative anticipating a potential disaster case and is now becoming a standard exercise in any business. In fact, many companies take ISO for BCP which is sometimes a prerequisite condition to make a business contract.

Flood control works by structural means increase safety level of a protected area which enables higher-level economic development to take place. Since no structural means can protect 100% from natural hazards, once unexpected hazards that exceed the safety level or unexpected accidents occur, the resultant disaster damage would be higher than the case before the increase of safety. Thus the higher the protection level by structural means, the higher the damage potential in the protected area. In other words, the physical flood control and damage potential are an endless race, which is sometimes called a vicious cycle. But this is an incorrect haste view. There are always possibilities of exposure control and reduction of vulnerability towards resilient society and sustainable development.

The vicious cycle argument often leads to an overemphasis of non-structural means over structural means. But the role of structural means should not be underestimated as it is absolutely necessary to secure an area in floodplain for wealth to be accumulated. Non-structural means such as early warning and evacuation can save lives but cannot protect properties and livelihood. Both structural means and non-structural means are necessary and should be integrated together under a best mix strategy.

5.5 INTEGRATED FLOOD RISK MANAGEMENT IN JAPAN

Given such reality, Japan came into the time of integrated flood risk management (IFRM) in the 1970s under the name of the Comprehensive Flood Control Measures policy. But it did not happen at once. IFRM is part of IWRM and Japanese integrated management of water resources has been progressed in a comprehensive way responding to various problems and needs emerging one after another. A brief chronology of judicial evolution shown in Tables 5.1 and 5.2 indicates such progress. This section will present IFRM following those laws and policies as they all contributed to form Japanese IFRM.

5.5.1 Overview of judicial actions related to IFRM

As explained in Table 5.1, the initial flood control policy was driven by the *Flood Control Triple Laws*, that is, 1st River Law, Sabo Law and Forest Law. It was the recognition of the necessity that flood control should start from mountains and forests to rivers. They are the basis of Japanese flood management reflecting the nature of the nation. Table 5.2 lists the follow-up laws and policies for IFRM within IWRM.

As for flood emergency management in local communities, the Flood Fighting Law of 1949 was enacted requesting local governments, that is, cities, towns and villages (CTVs) to take responsibility and form flood-fighting

Table 5.2 Representative Laws and Policies for IFRM in Japan.

Date	Law/Policy	Emphasis
1949	Flood Fighting Law (Showa 24 Law No. 193, 1949)	Flood fighting management association, community flood fighting teams, responsibility, fighting actions
1956	Industrial Water Law (Showa 31 Law No. 146, 1956)	Groundwater pumping regulation for industrial use
1957	Specific Multipurpose Dam Law (Showa 32 Law No. 35, 1957)	Promotion of flood control dams jointly with other users
1958	Landslides Prevention Law (Showa 33 Law No. 30, 1958)	Designation of landslide risk area, development regulation and prevention work
1962	Groundwater Pumping Regulation for Building Use Law (Showa 37 Law No. 100, 1962)	Groundwater pumping regulation for cooling and toilet use in buildings
1967	2nd River Law (Showa 39 Law No. 167, 1967)	See Table 5.1
1969	Steep Slope Collapse Prevention Law (Showa 44 Law No. 57, 1969)	Designation of steep slope risk area, development regulation and prevention work
1970	Water pollution Control Law (Showa 45 Law No. 138, 1970)	Effluent water quality control
1977.6	Comprehensive Flood Control Measures (River Council Interim Report, 1977)	IFRM in rapidly urbanizing area
1987.3	High-standard levees (River Council Report, 1987)	Promotion of super levees against exceeding floods
1997.6.12	3rd River Law (Heisei 9 Law No. 69, 1997)	See Table 5.1
1997.12.25	Environmental Impact Assessment Law (Heisei 9 Law No. 81, 1997)	Compulsory EIA and countermeasures in development
2000.5.8	Sedimentary Disaster Prevention Law (Heisei 12 Law No. 57, 2000)	Soft measures for managing sedimentary disasters, hazard maps designation of yellow/red zones, evacuation, training, relocation

Date	Law/Policy	Emphasis
2000.5.19	Special Law on Public Use of Deep Underground (Heisei 12 Law No. 87, 2000)	Promotion of deep underground for rivers, cities, traffic, lifelines, etc.
2000.12.19	Effective Flood Management Including Basin Resistance to Floods (River Council Interim Report, 2000a)	IFRM all over Japan
2015.1	Disaster prevention and mitigation responding to the new stage (MLIT special committee recommendation, 2015)	Prepare for the worst-case scenario. Save human lives as the 1st priority.

Dates indicate the promulgation day. References are given in parentheses ().

management associations under which community flood-fighting teams are formed. Such local system has been operating now with increasing importance as will be explained in Section 6.2.1.

In the 1950s, the economic recovery started and soon necessitated to stop land subsidence by groundwater pumping for industrial water use. The Industrial Water Law of 1956 was enacted to stop groundwater pumping for industrial water use. Land subsidence in Tokyo and Osaka started before World War II and was the major cause of flooding in urbanized areas. This was extended to the Groundwater Pumping Regulation for Building Use Law in 1962 to stop cooling and flush-toilet water use in buildings. By those regulations, land subsidence stopped in most cities by the 1970s (Environmental Agency, 2000).

Specific Multipurpose Dam Law of 1957 was enacted to promote a unified water use and control of river basin by constructing dam reservoirs. A dam should be constructed for flood control together with environmental flow (to maintain normal river flow for environment), irrigation, hydropower and municipal water supply including industrial water. Dam construction is led by and owned by the Ministry of Construction (now MLIT) inviting and sharing costs and benefits with other users (Senga et al., 1981). This idea was originally proposed by MONONOBE Nagaho in 1926 (Yasui, 2011) and River Water Regulation Project started in 1935 considering TVA type comprehensive river basin development. By this law, many multipurpose dams (526 as of 2019) were constructed for flood control and other purposes (MLIT, 2019b). The cost allocation method was specified as a separable cost-alternative cost-appropriate expenditure method in this law.

For increasing number of landslides and debris flow disasters in urbanizing areas, Landslides Prevention Law of 1958 was enacted to promote countermeasures for landslide risk areas including heap of coal wastes and prevent actions to increase risk such as cutting or filling soils or increasing subsurface water. This was followed by the Steep Slope Collapse Prevention

Law of 1969 that apply for the area over 30 degrees in gradient to promote protection works and restrict development similar to the Landslides Prevention Law. These two laws and the Sabo Law to stop mountainsides collapse in the Meiji Era 1897 are called *Sabo Triple Laws*. They are basically hazard control for sedimentary disasters. It was necessary to wait until the year 2000 to have soft management legislated by Sedimentary Disaster Prevention Law for such actions as provision of hazard maps, designation of risk area (yellow zone) and high-risk area (red zone), early warning, evacuation support for vulnerable people, compulsory regular training for yellow zone residents, development regulation of red zone, relocation support for the people living in red zone, etc. As of 30 September 2021, there were 669,219 yellow zones and 565,305 red zones designated (MLIT, 2021c).

For water quality control for rivers, lakes and groundwater, effluent regulation from industry was enacted as the Water Pollution Control Law of 1970 against poisonous materials, BOD etc. This reflected the facts that during the rapid economic growth in the late 1950s to 1960s, many urban rivers were deadly contaminated in black with intolerable odor and tragic heavy metal diseases such as Minamata disease in Kumamoto Prefecture and the 2nd Minamata disease in the downstream of the Agano River in Niigata Prefecture by Methyl Mercury used as catalyst to produce aceto-aldehyde, and Itai-itai disease in Toyama Prefecture in rice contaminated in the Jintsu River by Cadmium discharged from Kamioka Mine (Oda, 2008). By this law with groundwater regulation, reuse of Japanese industrial freshwater was promoted and by 1975 the reuse rate was already over 2/3 and now 79% for factories over 30 laborers (MLIT, 2022).

5.5.2 Comprehensive Flood Control Measures (1977)

Now, about Comprehensive Flood Control Measures, the Japanese IFRM policy. In June 1977, the River Council of the Ministry of Construction (MC, now MLIT) issued an interim report on Comprehensive Flood Control Measures (River Council Interim Report, 1977) which was a recommendation responding to the inquiry from the government on necessary flood control measures against increasing urban floods. It was a new flood management policy for rapidly urbanizing basins. The strategy was to extend flood management measures from riverine control such as dams, canals, levees, river improvement works to source control such as infiltration and storage in residential areas as well as non-structural human adjustment means such as early warning, evacuation and other preparedness measures. SAITO Naohisa, then an officer of the MC in charge of urban rivers described the policy as follows (Saito, 1985):

1. To identify the storage and retardation capacity that a concerned river basin ought to have together with riverine capacity and design a plan to achieve it.
2. To identify and publicize the anticipated flood inundation and debris flow risk areas.
3. To identify the short-term urgent targets of flood control infrastructure development instead of just setting a long-term goal.
4. To make a land use plan and an architecture design plan in the basin to respond to the current state of flood safety and the urgent protection targets.
5. To strengthen the system of prompt information dissemination for residents at time of floods.
6. To establish preparedness, warning and evacuation system in debris flow risk areas.
7. To strengthen local flood-fighting system.

Responding to this policy recommendation, an inter-bureau committee was organized in the MC across planning, urban, road, housing and river bureaus, and their independent projects were integrated into a coordinated set of comprehensive flood control measures to mitigate urban floods in rapidly urbanizing basins (Saito, 1985). Namely, in 1979, they designated special rivers for comprehensive flood control measures, first 9 rivers and eventually extended to 17 rivers by 1988, mostly in Central Japan and one in Hokkaido. They were selected from rapidly urbanizing rivers with catchment area of 30–1000 km², population density more than twice since 1955 or over 1000 persons/km², urbanized areas more than 20% and increasing or already over 50%, safety level less than 50 mm/h rainfall or 1/10 probability (MLIT, 2003b).

In 1980, the MC Under-Secretary order was issued to request the designated rivers form a *basin coordination committee* to integrate basin management on rivers, roads, cities, sewerage, land use planning, etc. by gathering respective officers in charge from national, prefectural and city, town or villages, and design a *basin management plan* to meet the flood safety against 50 mm/h heavy rains of the 10-year return period. Such a plan included not only riverine facilities but also basin capacities (MLIT, 2003b). Detailed explanations are available in MLIT materials for program assessment on Comprehensive Flood Control Measures (MLIT, 2004).

A unique strategy of this plan was the allocation of discharge to riverine and basin for them to share responsibility for safe treatment such as 80% and 20% of discharge. Riverine measures to treat its portion are such as dams, levees, diversion canals, channel improvements, detention ponds or tunnels in or along rivers. Basin measures are classified into three according to different characteristics of basin areas. One is so-called *storage area* such as forests and other natural land where urbanization should be controlled to conserve them. Urbanized areas should also serve as storage area by constructing and installing artificial storage and infiltration facilities such as disaster adjusting ponds, rain harvesting cisterns, infiltration pavement, infiltration pits and pipes, etc. The second is so-called *retardation area* where rainwater is temporarily retarded such as rice paddies and other depressions which should be conserved as much as possible from urbanization. The third is the *lowlands* where residential houses and districts themselves should be protected from floods by flood-proof walls or gates, pilotis, pumps to drain inland rain, etc. (MLIT, 2003c).

The target basins were urban rivers which were defined as rivers flowing urban area with densely inhabited district (DID) population over 30 thousand or development of residential area over 100 ha and classified into three types: *urbanizing* basins like the Tsurumi River basin, *urbanized* basins like the Kanda River and the Meguro River basins, and *zerometer zones* like the Sumida River basin (Saito, 1985). Formation of multi-stakeholders' basin coordination committee was compulsory to jointly design flood management plan both structural and non-structural over rivers and a basin. It was, in fact, Japanese exercise of IWRM for urban floods to integrate management of forests, plains, urbanized areas, river courses, sewage systems, precipitation observation, flood forecasting, early warning, hazard maps, preparedness, all-inclusive. It truly considered "comprehensive" measures.

Background

The initiation of this policy reflected the occurrence of a number of serious urban floods since the late 1950s especially in rapidly urbanizing basins such as, in the early stage, the Neya River basin in Osaka, the Tsurumi River basin in Kanagawa, etc. Urbanization is a result of population concentration, economic development and livelihood and assets accumulation, and in such a basin the water storage and retardation function decreased. Urban flood problems were so serious and even quite a few lawsuits were filed. The first case was "the Daitoh flood disaster suit" filed in the Osaka Regional Court in 1973. The Yata River in Daitoh City is a small branch of the Neya River where flood inundation occurred by overflow on 12–14 July 1972. The residents sued the governments claiming their defect in the administration of the river. It took 17.5 years to have been finally concluded with the loss of the residents' side at the Supreme Court in 1990 (Taniguchi, 2006). The Supreme Court remanded the original residents' victory in 1984, which started so-called a winter period of the lawsuits by residents against governments.

An exception was the case of the Tama river where the residents won in 1992 after 16 years of endeavor. The Tama River flood occurred on 1–5 September 1974 caused by heavy rains stimulated by Typhoon 16. The flood eroded the Tama River levees at Komae for 260 m and flew 19 houses into the river as shown in Figure 5.13. It was finally over by the explosion of the

Figure 5.13 The dike breach site at Komae City, Tokyo by the flood of Typhoon 16, 1974. Photo was taken just after the flood is over. The residential area carved out and the dynamite broken central part of the weir are seen. (Photo taken by Komae City on 6 September 1974, Courtesy of Komae City, Tokyo)

weir by dynamites that was directing flood flow to breach the levee (Komae City, 1975; Komae City: website on the Nightmare of the Tama River Bank Breach; Tamagawa Sampo: website on the Monument of the Tama River Breach). The residents had sued the governments in 1976 and they won in the Tokyo Regional Court but lost in the Tokyo High Court. But Supreme Court admitted the government's management defect and remanded the Tokyo High Court judgement. Then the residents finally won the suit in December 1992 (Miyoshi, 2015).

On 12 September 1976, the Nagara River levee breached for 50 m at Anpachi and inundated Anpachi and Sunomata towns in Gifu Prefecture by the flood caused by heavy rains brought by Typhoon 17 (Gifu Prefecture:

website on Heavy Rain Disaster on 12 September 1976). Thousands of houses were inundated and the residents sued the governments in 1977 but lost at the Supreme Court in 1994, so as many other cases (Taniguchi, 2006).

Including the concern for lawsuits, the urban floods became the major concern of the government.

Tsurumi river experiences

The Tsurumi River basin (235 km²) located in between Tokyo and Yokohama is considered as a forerunner of the comprehensive flood control measures because it faced the problem of urbanization and urban floods from the early stage of Japanese economic growth and started the use of comprehensive measures ahead of any other basins in Japan. A rapid urbanization of the basin was remarkable. Population 450,000 in 1958 tripled to 1.2 million in 1975 and then 1.84 million in 2000 as shown in Figure 5.14. During this growth, they suffered from a series of floods such as of 1958.9 (Kanogawa Typhoon), 1966.6 (Typhoon 4), 1976.9 (Typhoon 17). They were urban floods caused by people's settling in the area used to serve for infiltration or retardation of water. The transformation of land use from paddy fields, forests, marshes or open fields that infiltrate rainwater or retard flood discharges to paved roads, house or building roofs and compacted land that quickly drain water to rivers made flood discharges from the area with higher peak and shorter concentration time as shown in Figure 5.15.

They basically took two approaches. The first one was to introduce, in addition to traditional riverine works such as levees, river improvement, riverbed dredging, retardation reservoirs, etc., basin facilities such as off-site retardation ponds and infiltration facilities to slow the speed of rainwater discharge to rivers when new residential areas were developed.

Figure 5.14 Progression of urbanization in the Tsurumi River basin, 235 km² located between Tokyo and Kanagawa prefecture. Pink part is the urbanized area. Population and urbanization rates were roughly (0.45M, 10%), (1.20M, 60%) and (1.84M, 85%) in 1958, 1975 and 2000, respectively. (Courtesy of MLIT: website on the Tsurumi River)

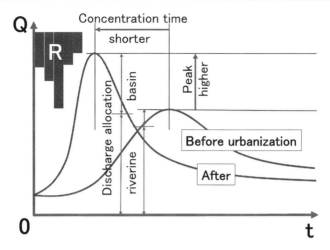

Figure 5.15 Impact of urbanization on flood discharge on its peak height and concentration time and the discharge allocation to be made for riverine and basin where most of discharge increased should be managed by basin rather than river.

This was mainly implemented by municipalities and the Japan Housing Corporation that started as early as the 1960s after the devastating flood disasters by the Kanogawa Typhoon in 1958 (Ando et al., 1976). The other was to establish the Tsurumi River Basin Water Disaster Prevention Committee in 1976 made up of the river administrator, local governments and communities, and intellectuals. The Tsurumi River is a small but transboundary river shared by two prefectures, Kanagawa and Tokyo within which four cities, Yokohama, Kawasaki, Machida and Inagi are included. Furthermore, administrative sections of rivers, housing, parks, sewerage, schools, etc. need to cooperate. The formation of such a basin committee, therefore, was critically important to reduce the flood risk of the basin and the Tsurumi River intersectoral committee established was the first of its kind in Japan.

The introduction of the basin facilities together with riverine efforts could reduce the flood damages to a considerable extent already in the mid-1970s in the Tsurumi River basin (MLIT: website on the Tsurumi River). In addition, since a river basin planning committee was formed to make transboundary and sectoral cooperation possible, its experiences became the model of the river council interim report of 1977 and the governmental authorization of comprehensive flood control measures as the national policy. Naturally, the Tsurumi River basin became the first designated river to apply the comprehensive flood control policy in 1979 (MLIT KTR et al., 2007).

In-situ storage and infiltration

The basic principle of hard measures in this policy was that land development creates no new flood discharge. The methodology promoted was storage and infiltration. As mentioned earlier and shown in Figure 5.15, land development makes flood peaks higher and arrival time faster because land development decreases slow discharge components and increases quick discharge components of rainwater. Before development, large portion of rainfall is stored in forests, ponds, rice paddies, grasslands, etc. and then infiltrates into underground and slowly discharges to land surface and rivers. But by land development, such slow discharging area is converted into residential, commercial, industrial or road areas where little rainwater is stored or infiltrates underground as land surface is covered by roofs, compacted land, pavements and other impermeable or less permeable surfaces for rainwater to quickly flow into diches, conduits and rivers resulting floods faster and higher. Promotion of storage and infiltration is to mitigate such impacts by a source control which is more fundamental for reducing flood peaks by basin itself rather than riverine facilities.

Figure 5.16 shows the rainwater infiltration type, often called discharge reduction type sewerage system designed and exercised soon after the policy of Comprehensive Flood Control Measures started in 1977. Since the Japanese sewerage system was then a combined type in general, in this system the rainwater was infiltrated and retarded as much as possible from holes of pipes, gravels of pit bases and percolation trenches before joining to the public sewerage. Figure 5.16(a) shows example system that was implemented in many apartment complexes by Housing and Urban Development Corporation in the 1980s (Nagasawa, 1980) and Figure 5.16(b) shows an example of storage and infiltration type system experimented by Bureau of Sewerage, Tokyo Metropolitan Government (Koyama and Fujita, 1984). Note that in Figure 5.16(a), the vertical pipes to directly infiltrate rainwater near to the groundwater level was not exercised to avoid the risk of groundwater contamination. In Figure 5.16(b), detours and storage-type manholes aimed to retard rainwater to flow out but did not become popular as too costly and easy to be filled by sediments (Fujita, personal communication, February 2022). Such experiments, however, helped the effective use of in-situ storage and infiltration facilities.

For in-situ storage facilities, residents installed rain harvesting tanks underground or basement reservoirs under big buildings and used stored water for flushing toilets, firefighting, car washing, garden/road sprinkling, etc. as so-called gray water supply without purification treatment. In some cases, tennis courts, elementary school playgrounds, parks were used for dual function to temporally store rainwater during heavy rains. Elementary school playgrounds' use for such purpose is considered meaningful for educational purpose as many pupils would experience such flooding once

(a)

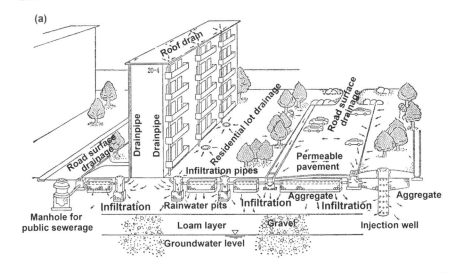

(b)

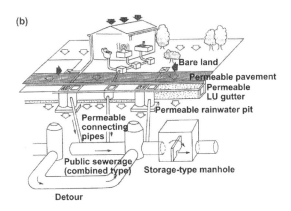

Figure 5.16 Examples of in-situ infiltration facilities. (Top) An example design of apartment complex developed by Housing and Urban Development Corporation in the early 1980s (Nagasawa, 1980). Rainwater fell impermeable surface is first infiltrated underground as much as possible by holes in pipes, gravels under pit bases, or infiltration trenches before it goes to urban storm drainage or the public sewerage. (Bottom) An experimental design of a residence by the Bureau of Sewerage, Metropolitan Tokyo. Domestic sewer directly goes to the public sewerage but rainwater first gets infiltrated or to urban rainwater drainage before it goes to the public sewerage (Koyama and Fujita, 1984). (Courtesy of NAGASAWA Yasuyuki and FUJITA Shoichi. Redrawn by FUJITA Shoichi for this publication.)

or twice in their elementary school age and trained for preparedness. Other than such dual purposes, many artificial retardation ponds were constructed nearby apartment complexes if there is suitable land available. Such off-site retardation ponds have difficulty to manage sediments, overflows, etc. similar to any other dam reservoirs.

As urbanization expands and intensifies, less and less urban rainfall is infiltrated and groundwater recharge decreases which results in less and less groundwater discharge through fountains and small rivers. Quite a few fountains and origins of small streams in urban area dried up destroying precious nature, fish, insects, grasses and flowers which was a serious degradation of urban amenity, too. It is a phenomenon of *urban desert*. Promotion of urban storage and infiltration serves as rainwater harvesting, recovery of groundwater wells, rich biodiversity and watery landscape back. Thus it is often called an *urban oasis* concept to secure urban water and environment (Musiake et al., 1987).

Suppose Metropolitan Tokyo has about 600 km² impermeable area where about 1500 mm rainfalls and 2/3 discharges, it is 0.6 km³ of water which is directly drained to the sea without any use. It is a large amount of waste of water resources. Meanwhile, Tokyo is getting water from upstream prefectures where considerable environmental destruction is enforced. If its own water resources are used in urban areas, its water security would much improve, and the environment of the surrounding prefectures would be saved. Urban rainwater storage and infiltration are not just for flood risk reduction but serve for precious nature back to urban area and the surrounding.

Establishment of FRICS

The other part of the recommendation of the interim report on comprehensive flood control measures was early warning and preparedness that promoted rainfall observations, forecasting and dissemination of information. In order to make it realized, the Ministry of Construction (now MLIT) established the Foundation of River & Basin Integrated Communications (FRICS) in 1985 and let them take charge of rain-gauging radars and disseminate the real-time rain clouds movements to all local flood management offices regardless of national or prefectural or municipal as long as in charge of flood management. The first C-band radar rain gauge was installed at Mt. Akagi in Gunma Prefecture in 1976. By 2019, 26 C-band radars (37–77 mm wave-length micro-wave, 1 km mesh resolution, 120 km range, transmitting every 5 minutes) and 39 X-band radars (25–37 mm wave-length, 250 m mesh, 60 km range, transmitting every 1 minute) radars were installed covering all over Japan as shown in Figure 5.17 (FRICS: website on Radar Rain Gauges). Since 2016, their combined rainfall data in 1 km mesh called XRAIN has been distributed every 1 minute.

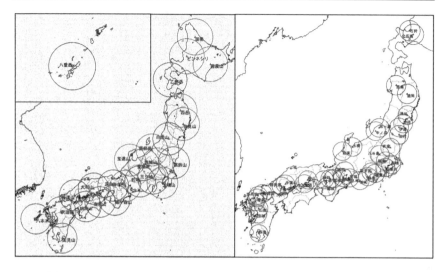

Figure 5.17 FRICS operates (left) 26 C-band and (right) 39 X-band radar rain-gauges as of 2021. Their combined XRAIN data in 1 km mesh every 1 minute has been distributed since 2016. (Courtesy of FRICS: website on Radar Rain Gauges)

5.5.3 Flood hazard maps

Another recommendation that the comprehensive flood control method proposed was preparation and dissemination of hazard maps. In the late 1970s, flood hazard maps were not popular in production and dissemination to local residents in flood-prone areas. The first flood hazard map produced in Japan is considered the work of OYA Masahiko in the form "Water-disaster morphology classification map of Nobi Plain, Kiso River Basin" back in 1956 (Oya, 1956). This map is basically a geomorphological map showing the classified topography and soil types of the area affected by repetition of floods in the past. This Nobi Plain geomorphological map was proven useful to indicate the area affected by the flood of Ise-Bay Typhoon in 1959 which brought the severest storm surge and floods in modern Japanese history as discussed in Section 5.3. Since then, the water-disaster morphology classification maps were produced in many river basins in Japan and eventually in various countries (NIED website).

Flood hazard maps were not formalized but historical flood information was tried to be publicized in some basins for preparation against floods. It was on the occasion of flood in the Abukuma river in August 1998 when the first investigation was made to identify the effect of availability of flood hazard maps on the citizens' evacuation behavior in Koriyama City, Fukushima Prefecture by KATADA Toshitake (1999) who found that the people who had seen the flood hazard map before the flood had evacuated, on average, about an hour earlier than the people who had not.

With this evidence, MLIT moved to seriously consider the use of flood hazard maps in disaster management operation and the government made the amendment to the Flood Fighting Law in June 2001 (Heisei 13 Law No. 46, 2001) to make use of flood hazard maps compulsory in so-called flood forecasting rivers (as of 2020.7.31, 298 minister designated rivers and 129 prefectural governor designated rivers that have large catchment area having enough forecasting time and potential to cause serious losses to national economy). In 2005 this was extended to so-called water level warning rivers (as of 2020.7.31, 150 minister designated rivers and 1560 prefectural governor designated rivers that are too small to take forecasting time but still have potential to cause serious losses to national economy and assigned to issue the water level warning when river has reached to the inundation risk level) (MLIT: website on Potential Inundation Areas and Hazard Maps).

In 2005, MLIT issued the "Flood Hazard Maps Producing Manual" (MLIT, 2005c) which said hazard maps should be prepared mainly by CTV governors indicating not only potential flood inundation areas but also evacuation information. Therefore, it is not just a potential hazard indication map such as by dike breach or overflow calculated by some experts but a guidance map indicating suggested evacuation shelters or places considering possible evacuation routes which should be decided by community members themselves through participatory approach.

The specific floods to be considered for calculating inundation areas for hazard maps in each river basin were specified in the "Basic Plan for Implementation of Construction Works" of each basin before 1997 according to the River Law of 1964. After 1997, the River Law of 1997 changed the "Basic Plan" above into the "River Management Basic Policy" and "River Management Implementation Plan". As mentioned in Section 5.4, the difference between the 1964 and the 1997 river laws was in their decision-making process with or without inviting comments from local residents before deciding on the plan.

Figures 5.18 and 19 show an old and a current hazard map of part of the Middle Tone River and the Edo River. Figure 5.18 is an example of an early one in 1997 which was a flood inundation risk area map and not exactly a hazard map yet as it did not include evacuation information. It was simply a map showing historical inundation area at the time of the Kathleen Typhoon flood in 1947 when the Tone River basin received 318 mm rainfall in 3 days that was considered once 200 years probability of occurrence. Here red indicates inundation depth over 2 m, yellow 0.5–2 m and blue below 0.5 m.

Flood hazard maps are improved a lot since then covering landslides and debris flows, storm surges and tsunamis as well for any communities all over Japan accessible from MLIT disaster portal site (Disaster Portal: website on Hazard Maps to Be Overlaid). The evacuation centers are also indicated at different places for different hazards. Figure 5.19 is the current flood hazard map

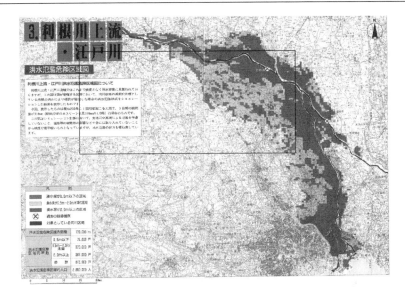

Figure 5.18 An early-stage flood hazard map (flood inundation risk area map) prepared by the Ministry of Construction in 1997 for the Middle Tone River and the Edo River areas based on the Kathleen Typhoon flood in 1947 (FRICS, 1997). (Courtesy of MLIT)

Figure 5.19 The current flood hazard map issued by MLIT for the Middle Tone River and the Ara River areas corresponding to the rectangular area of Figure 5.18 under an anticipated maximum scale precipitation (AMSP). The map indicates water depth in yellow to pink color and evacuation centers by persons that were designated by local governments. In this area, only flood is the potential hazard but in other areas sedimentary, storm surge and tsunami hazard maps can be overlaid if they are subject to risk. (Disaster Portal: website on Hazard Maps to Be Overlaid)

of the rectangular area indicated in Figure 5.18. Since the scale is small, the evacuation centers are many. The inundation depth is shown wherever the cursor is clicked. The inundation area and the depth correspond to the anticipated maximum scale precipitation (AMSP) which is considered about or more than 1000-year return period (Takeuchi and Tanaka, 2021) responding to the "new stage recommendation" in 2015 to be discussed in Section 5.5.7. In nationwide, maps are in some regions still under revision to the ones based on the ASMP.

5.5.4 High-standard (super) levees

The River Council of the MC submitted its recommendation on high-standard levee to the government in March 1987. It is an idea of a levee that never breaches even by overtopping. As shown in Figure 5.20, the width of the levee (W) is suggested to be about 30 times more than the height of the levee (H), that is, $W \gtrsim 30H$ by which if river flow overtops, water slowly flows down on the back side of the levee and the levee never breaches as it is so wide. It was immediately implemented in 6 major rivers in 1987 to protect highly populated areas, i.e., the Tone, the Edo, the Ara, the Tama, the Yodo and the Yamato Rivers for 873 km. The first section of 1.78 km was completed in the Tone River at Yako, Chiba Prefecture in 1992 (Chiba

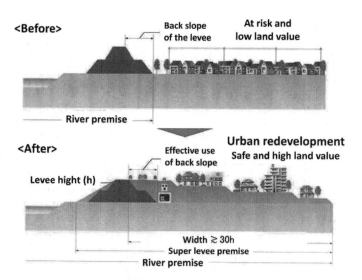

Figure 5.20 An example of urban redevelopment by a super levee. The width of the levee is greater than 30 times of the levee height. The cost may be recovered by the increase of land price due to increase of safety. (MLIT: website on High Standard Levees)

Prefecture: website on Super Levee at Yako). Note that Metropolitan Tokyo also has had a similar project since 1985 in the Sumida River but its spec is W up to 50 m, smaller than that of MLIT (Metropolitan Tokyo, 2016).

The progress was very slow due to lack of funds. The construction was once halted in October 2010 and reopened in December 2011, having reduced the target length to 120 km only in the zero meter zones in 5 rivers, the Edo, the Ara and the Tama Rivers in Tokyo and the Yodo and the Yamato Rivers in Osaka. But still as of March 2017, the completed were 14.3 km (12%) where the section with the full super levee size was only 3.3 km (2.8%) (MLIT: website on the Current Status of Super Levees).

Construction of super levee is very costly and time-consuming. It is considered that the cost can be recovered by the increased land value by making the area free from flood hazards and useful for value concentrated activities for residents, business, transportation, etc. But the difficulty is extraordinary. It takes a long time from conceptualization to implementation and real use. The discussion on the plan to agree among all stakeholders including the project implementation authority, landowners, residents, apartment owners to provide temporary housing, etc. would be the most difficult part. It is also not an easy matter for residents to move out for a few years during construction and keep their livelihood and family life elsewhere. Super levees are considered suited for construction as part of urban redevelopment project. Some alternative technology such as double steel sheet piles or steel pipe piles method (or implant method) is proposed and examined in sea levees (walls) in Kochi Prefecture which use continuous steel piles in levees making them resistible against breach even under earthquake or tsunami (Japan Bosai Platform: website on Implant Method).

5.5.5 Underground rivers

Underground rivers for storage and drainage of floodwaters were another way to manage excess floods in urban area. In the 1950s–1970s the population increase in large cities was so rapid and intense that there was no room left for enlarging river sections to increase discharge capacity of urban rivers which necessarily made river managers to seek space underground.

The Neya River South and North in Osaka Prefecture

The first underground river in Japan was planned and constructed in Osaka Prefecture in the Neya River basin where 3/4 basin was below riverbeds and needed pumping-up to drain rainfall to rivers as Figure 5.21 describes. The Neya River basin is 267.6 km² area located in the central part of Osaka Prefecture that origins from the Ikoma mountain ranges in the East and joins to the Oo River (the old Yodo River) in the North West. The basin is a low-lying land, used to be part of the Osaka Bay till 6–7000

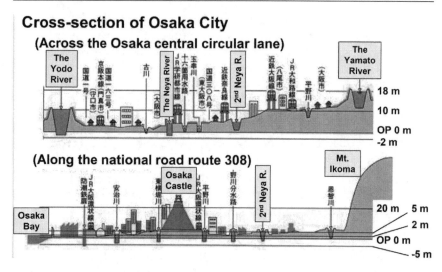

Figure 5.21 Cross-section of the central part of Osaka Prefecture where the Neya River basin is located. OP indicates the average sea level of Osaka Bay. (Osaka Prefecture: website on Comprehensive Flood Control Plan)

years ago, bounded by two sky rivers, the Yodo River in the North and the Yamato River in the South, and the Uemachi Plateau in the West (Osaka Prefecture, 2015). It has been often hit by floods. The largest one was the Daitoh flood in July 1972.

The first underground river project started in 1981 and the earliest section put to use was an underground storage tunnel in 1990 (Sankei News, 2019; Osaka Prefecture: website on Underground Rivers). It was only a disconnected portion of the underground river to temporarily store floodwater which would eventually be connected to a long underground river to drain the floodwater to the Oo River.

Now there are two underground river projects on-going, that is, the Neya River North Project and the Neya River South Project where the latter includes the earliest section completed as mentioned earlier. The North Project is 14.3 km long (completed 9.7 km by 2021) for 250 m³/s to drain by connecting the upstream of the Neya River to the Oo River, and the South Project is 13.4 km long (completed 11.2 km by 2006) for 180 m³/s to drain by connecting the 2nd Neya River (a branch in the mid of the basin) to the Kizu River near to the Osaka Bay. They aim to secure the storm water drainage up to 50–65 mm/h rainfall. Until their completion, the completed tunnels and shafts are used as flood retardation storage of capacity 200,000 m³ and 630,000 m³, respectively (Osaka Prefecture, 2015; Osaka Prefecture: website on Underground Rivers).

The 7th Circular Road of Metropolitan Tokyo

In Metropolitan Tokyo there has been a giant underground river plan since 1987 as recommended by the Committee on Conceptualization of Metropolitan Tokyo Underground River requested by the Governor of Tokyo in 1985 (Izumi, 1990). This plan is to use the underground of the 7th and the 8th Circular Roads of Metropolitan Tokyo to enable to complete the urban drainage capacity to 50 mm/h and eventually extend to 75 mm/h heavy rains by retarding floodwater of 10 rivers including the Shakujii River and the Kanda Rivers by 12.5 m diameter tunnel in 40–60 m deep underground. Eventually, the tunnel will be 30 km long and directly discharge floodwater to Tokyo Bay. The first 2 km floodwater retarding tunnel (240,000 m³) between the Zenpukuji and the Kanda Rivers and a large floodwater intake shaft from the Kanda River was started construction under the 7th Circular Road in 1988 and put to use in 1997. The second section, 2.4 km, 300,000 m³ tunnel between the Zenpukuji and the Myoshoji Rivers was completed for use in 2007 (Metropolitan Tokyo: website on the Underground Rivers Under the Kanda River and the 7th Circular road).

In addition to such underground tunnels and rivers, the Metropolitan Tokyo has been constructing other types of flood retardation facilities such as ponds along rivers and underground storage boxes that are also to upgrade the stormwater drainage capacity for 75 mm/h rainfall or, if not yet achieved, 50 mm/h rainfall (Metropolitan Tokyo: website on Flood Control Works of Small to Medium Scale Rivers).

The Naka River and the Ayase River Basin

The Naka River and the Ayase River (a branch of the Naka River) basin is a low-lying A-class river basin located in the north of Tokyo having cities such as Kasukabe, Soka, Yashio, Iwatsuki, etc. They were hit by serious floods a number of times since Typhoon Kathleen in 1947 but became the bed town of Tokyo especially since the mid-1950s. In this basin, there are 5 rivers, the Ayase, the Moto-Ara, the Oo-otoshi-Furutone, the Kuramatsu and the Naka Rivers in parallel from the West to the East, and at the East end, they are bounded by the well-protected Edo River (Figure 5.22). Various flood control works were implemented in the basin to raise flood safety. The latest most unique one was to connect the Oo-otoshi-Furutone, the Kuramatsu, the Naka Rivers and other small rivers in between by a 50 m deep, 10 m diameter underground river to collect and store river water and dump it into the Edo River by giant (14 m head and the total capacity 200 m³/s) pumps in a vertical 70 m deep cylinder hole often called "underground shrine". Figure 5.22 shows the total image of the system and Figure 5.23 a cross-section of the pipe and the storage chamber from where floodwater is pumped out to the Edo River after its flood. This 6.3 km long system is the Metropolitan

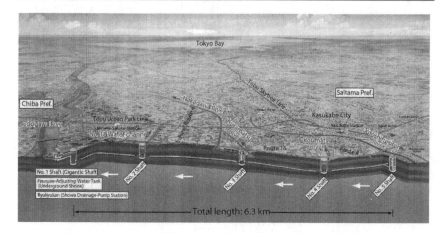

Figure 5.22 The total system of the Metropolitan Area Outer Underground Discharge Channel. (Courtesy of MLIT, Edo River Office: website on the Metropolitan Area Outer Underground Discharge Channel)

Figure 5.23 The lower left corner is the underground river with 10 m in diameter. The main picture shows the large chamber with tall columns (often called underground shrine) where floodwater is stored until pumped out to the Edo River when flood is over. In Osaka, even longer underground rivers are constructed in the Neya River basin. (Courtesy of MLIT, Edo River Office: website on the Metropolitan Area Outer Underground Discharge Channel)

Area Outer Underground Discharge Channel, for which the project started in 1992 and completed for use in 2012. It costed ¥235 billion of which about 60% was considered already recovered by October 2019 by avoiding the potential flood losses in the beneficiary urban areas (MLIT, 2012a; MLIT Edo River Office: website on the Work of the Office).

Now the construction of underground rivers is underpinned by Special Law on Public Use of Deep Underground (Heisei 12 Law No. 87, 2000) enacted in May 2000 which provided a judicial framework for the public use of deep underground space for various purposes. This act aimed to respond to the difficulty of developing new access to the central part of urban area, such as for railways, traffic roads and river courses. Here the deep underground is defined as the underground space below the ordinary use for basement (40 m) or foundation (supporting layer stronger than 2500 kN/m) plus safety distance (10 m) (MLIT, 2018). Under this jurisdiction, various underground river projects were revitalized.

5.5.6 Effective flood control including basin resistance

The last recommendation of the River Council of the MC in the 20th century was the interim recommendation on "Effective Flood Control Including Basin Resistance" on 19 December 2000 (River Council Interim Report, 2000a). It was an extension of the established Comprehensive Flood Control Measures in rapidly developing basins issued in June 1977 to any basins in Japan. But this recommendation had a special significance that was not in the 1977 policy, namely, it was an official admittance of the notion that the conventional flood control measures alone cannot keep flooding out of residential areas. This reflected an increasing frequency and severity of flood damages due to heavy rains under increasing societal vulnerability and climate change. But it was not a denial of the conventional methods but only an addition of "including basin resistance" to conventional means.

The conventional methods were to confine discharge into river courses and flow it out to the sea as soon as possible. This was the principal methodology of flood control since the Meiji Era and the spirit of the first River Law. Under this principle, many riverine facilities such as continuous levees, diversion canals, dam reservoirs, etc. had been constructed all over Japan. But it was revealed that it was impossible to manage floods by such riverine methodology alone. It is necessary to manage them from their source, that is, the basin as a whole where discharge occurs. This was the original idea of Comprehensive Flood Control Measures already put forth in 1977 in rapidly urbanizing basins such as the Tsurumi, the Kanda and the Neya River basins. The new millennium introduction of this policy extended its application to any basins in Japan including rural areas under the current social and climatic changes. At the same time, it admitted the necessity of preparing for the cases that floodwater come to residential areas and to take actions for mitigating damages.

The basic concept of the policy was first, any basin is as shown in Figure 5.24 divided into four areas 1) flow generating area, 2) flooding area in a narrow floodplain, 3) flooding area in an extensive floodplain and 4) and urban protection area. Then, each area was assigned to share the roles for flood control as follows:

1) Flow generating area: This is an area where discharge is generated. It includes mountains, hill slopes, forests and other uninhabited lands where discharge occurrence should be made as much as delayed, and controlled especially when new developments come in. The methodologies for such discharge control include keeping forests' water holding capacity, maintaining other retardation storages, constructing dam reservoirs and their effective operation to delay and decrease discharge.

2) Narrow floodplain: This is an area typically in upstream mountainous area where floods do not expand to wide areas and continuous levees are not reasonable as they occupy too much scarce land resources of the area. The methodologies suggested are such as land elevation and ring dikes to protect only the necessary properties and zoning for land use control considering flood risk.

3) Extensive floodplain: This is an area typically in middle to downstream basin or plain where floods may propagate and expand to large area that includes paddy fields, farmlands, scattered residential areas and industrial areas. The methodologies suggested are such as maintenance of open dikes to reduce downstream floods by temporarily inundating upstream and draining it after flood, secondary dikes to stop further expansion of floods and floodproofing of each house or a block of houses. Of course, since such options allow inundation of land, selection of inundation area should be carefully made considering the characteristics of land use.

4) Urban protection area: This is an area where many people and properties are concentrated. As flood vulnerability is high, careful countermeasures and preparation are necessary. The methodologies suggested include, among others, joint operation of rivers with sewerage, that is, river managers and sewerage managers should well collaborate on timing of pumping inland floods into rivers receiving outer floods, floodproofing of lifeline and underground facilities such as subways and underground shopping malls, hazard maps, warning and evacuation. In some cases such as densely populated zero meter zones, no levee breaches are allowed to occur and the construction of super levees may be considered along major rivers.

Once again it is an extension of the flood control means from riverine to the whole basin not only in specific rapidly urbanizing areas but also everywhere

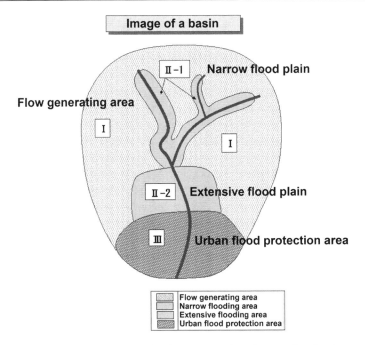

Figure 5.24 Schematic illustration of a basin area classification for flood control by basin, modified from River Council Interim Report (2000b). (Courtesy of MLIT)

in Japan. The necessity of such extension reflects increasing urbanization and vulnerability. It was especially important to note that this new policy allows inundation in some inhabited areas by accepting open dikes, secondary dikes, protection of houses by floodproofing, etc. For some traditional engineers who believed in engineering capability of complete control of floods, it was rather difficult to swallow the concept.

5.5.7 Disaster prevention and mitigation responding to the new stage

Recent linear rain bands and super typhoons are two major new types of meteorological extremes under climate change. Linear rain bands caused unprecedented sedimentary disasters by landslides hitting residential areas such as: On 29 June 1999 in Hiroshima (dead and missing 39 as a total in Japan from June 23 to July 3), on 11–14 June 2012 in Northern Kyushu (dead and missing 57), and on 20 August 2014 over Hiroshima

Prefecture (dead and missing 77, especially in Asa South and Asa North districts).

On 8 November 2013, Super Typhoon Haiyan hit Leyte Island, the Philippines which caused a tsunami-like storm surge and took the dead and missing more than 7000 lives and it was considered that something similar would hit Japan any time. In fact, in October 2019 equally powerful Super Typhoon 19 (Hagibis) hit East Japan and took 107 lives.

Such intensification of flood-related hazards was considered as a "new stage" of the climate–human relation and the government responded to it with a new policy recommended by a special committee on "Disaster prevention and mitigation responding to the new stage" in January 2015 (MLIT Special Committee Recommendation, 2015). The basic strategy proposed was to prepare considering the worst-case scenario which had been already accepted in earthquakes and tsunamis management after the Great East Japan Earthquake and Tsunami (GEJET) but not yet in flood-related disaster management.

What the committee recommended was a methodology to anticipate the maximum scale of heavy rain and prepare life-saving actions against it. It was pointed out that since structural means cannot protect all life and property against the worst-case rainfall, it is necessary to put highest priority to protect human lives and avoid catastrophic losses even in the case where great property damages are inevitable. In practice, a new precipitation standard called "Anticipated Maximum Scale Precipitation (AMSP)" was suggested for hazard mapping to avoid unanticipated events. The return period of the AMSP was selected about once 1000 years as the worst-case scenario.

The methodology of calculating AMSP was then issued by MLIT (2015) which instructed to use the precipitation records observed in the past by Japan Meteorological Agency (JMA) instead of the projection data simulated by various global circulation models (GCMs) or regional climate models (RCMs). It was because although many advanced climate change simulation models were available, their predictions were still far uncertain in time and space resolution to be used for practical hazard mapping for local communities' use. Meanwhile, the JMA observation records were long accumulated in high accuracy and high density in time and space. Everywhere in Japan is covered by about 1300 AMeDAS stations or about 17 km mesh at least since 1979. Still more, although the number is much less, there are stations with much longer observation records accumulated since the start of JMA observations in 1875.

The selected approach to estimate the AMSP was to use those trustable historical data to identify the historical maxima in each region (the whole of Japan was divided into mutually exclusive 15 regions) in terms of duration-area-depth (DAD) curves. The assumption used was to consider the identified maximum DAD in a region occurs anywhere in the region. The resultant precipitation was considered roughly 1/1000 probability of occurrence. The detailed procedure was described in the MLIT-issued guideline in 2015 (MLIT, 2015).

A critical review of this policy and the procedure is given in an article by Takeuchi and Tanaka (2021) that clearly expresses its rationale and limitation of its applicability in a historically well-observed data-rich nation with necessary revisions as climate change advances. Also, this must be an interim policy until detailed climate change projections become available in high resolution in high accuracy.

With a new revised hazard map, the government started issuing new early warning which was for residents to make timely evacuation before too late. The new guideline issued for evacuation instruction by the Disaster-in-Charge of the Cabinet Office in March 2019 was Table 5.3.

There are "heavy rain warning" and "heavy rain special warning". In Level 3, with "heavy rain warning", the vulnerable people who need support for evacuation such as patients and the elderly in the disaster anticipated area are requested to evacuate. In Level 4 with "evacuation instruction", all people in the disaster anticipated area are requested to evacuate. In Level 5 with the "heavy rain special warning", a heavy rain that have never been experienced before is happening and disasters would have already occurred in the target area so that evacuation is too late and no longer suggested, so that all people in the disaster anticipated areas are requested to take possible actions

Table 5.3 Warning Levels for Evacuation and Issuance of Warnings.

Warning Level	Necessary Action	City, Town, Village Issuance	JMA Issuance
5	Protect life (Too late to evacuate)	Disasters occurred	Heavy rain special warning. Inundation occurrence information
4	Evacuation	Evacuation instruction	Sedimentary disaster warning information, Flood risk information, Storm surge special warning Storm surge warning
3	Elderly and other vulnerable people: evacuation, Others: preparation for evacuation	Elderly and other vulnerable people start evacuation, Others start preparation	Heavy rain warning, Flood warning, Inundation warning information, Storm surge caution
2	Check for evacuation	-	Heavy rain caution, Flood caution warning, Inundation caution information
1	Mental preparation	-	Early caution information (could be warning level)

Evacuation should be completed during level 4 before Heavy rain special warning is issued in level 5. (JMA: website on Correspondence between the Warning Level and the Warning Information).

to protect their own life on site such as horizontal and vertical evacuations at home.

5.6 SUMMARY OF JAPANESE EXPERIENCES ON FLOOD-RELATED DISASTER RISK MANAGEMENT

Rather lengthy descriptions of Japanese experiences on flood-related disaster risk management were presented and it would be beneficial for readers to have a short summary of the experiences and lessons learnt.

5.6.1 Summary of experiences

Japanese history of managing flood-related disasters was an experience of living with floods. As society changes, the way of living with floods also changes, that is, an evolution of flood management. The modern flood management in the Meiji Era took the principle of keeping floodwater away from people by encroaching flood discharge within rivers by continuous levees and releasing it to the sea as soon as possible. This was a considerable change from the old-time methodology in the Edo Era when there were no resources nor technology to build continuous levees. They used to select higher places to live avoiding frequent visits of floods or construct minimum local protection such as ring dikes or open dikes to protect only targeted houses or part of villages necessary to keep out of floods.

Continuous levees were people's dream for a long time and in the Meiji Era, with the help of Dutch engineers, the dream was tried to be realized and continuous levees were constructed all over Japan and raised their height when floods exceeded. Also, many diversion canals were constructed such as the Yodo River in 1910 designed by OKINO Tadao, the Ara River in 1930, the Edo River in 1919, the Okozu Diversion Canal in 1927 directed by MIYAMOTO Takenosuke and AOYAMA Akira, and others.

Although the history of dam reservoirs for irrigation purpose goes back to the 7th century and for hydro-power generation in the Meiji Era, those for flood control started after the introduction of Specific Multipurpose Dam Law of 1957 (Showa 32 Law 35, 1957). Since then, many flood-control dam reservoirs were constructed.

Note that the oldest dam construction in Japan was the Sayama Pond (Osaka Prefecture) built around 600 (Kanamori et al., 1995; Osaka Sayama City Education Committee, 2014) and the Man-no Pond (Kagawa Prefecture) built originally in the Taiho Era (701–704) and, after a collapse, in 821 by a respectful Buddhist monk Kukai (Man-no Town Education Committee, 2008).

Under the confine and quickly drain to the sea principle, together with advancement of meteorological observation and dissemination of forecasts, Japan achieved a dramatic reduction of death tolls since the 1960s which was a Japanese success story. Nevertheless, the economic losses of flood and sedimentary disasters did not decrease. This was due to the increase of social vulnerability or flood damage potential especially during the period of rapid economic growth from the 1960s to 1980s. In urban area and isolated hillsides, floods and landslides and debris flows were difficult to be controlled and as a result, death tolls never get below several tens and economic losses stay huge.

In order to address such issues, it was realized that the traditional principle since the Meiji Era should change. What was put forward was a new policy "Comprehensive Flood Control Measures" issued in 1977 (River Council Interim Report, 1977). The main principle was "no new flood discharge by land development" by promoting in-situ storage and infiltration to retard rainwater where it fell as much as possible. It was in fact the same concept called as "water sensitive urban design" or "sponge city" that were introduced later in many countries such as Australia (Wong, 2006) and China (Peng and Reilly, 2021).

Such principle originally applied only to rapidly urbanizing basins but at the end of the 20th century extended to any basins in Japan reflecting recent societal as well as climate changes (River Council Interim Report, 2000a). At the same time, it was declared that the traditional strategy to keep floodwater out of human settlements by continuous levees, dams and other structural means was no longer almighty and methodologies to accept floodwater to come to residential area should be added where structural means alone could not work enough. In fact, against any hazards, there is no way to eliminate risk to zero so that mitigation methods should be considered and prepared to make damages small whenever disasters occur.

It is now in a "new stage" of flood management under climate change (MLIT Special Committee Recommendation, 2015). Japan is facing heavy rains that have never been experienced before. The linear rain bands are new form of heavy rains and super typhoons that affect large part of Japan are another type. Under such meteorological extremes, the target objective of flood management is no longer no economic losses but "no deaths" allowing flood damages to an inevitable level. In order to minimize losses and avoid loss of lives, the keys are preparedness, early warning and evacuation. This is not at all a matter of public responsibility alone but individuals and communities together. In other words, collaboration of individual help, mutual help and public help supported by high-quality regionalized and targeted hydro-meteorological forecasts are indispensable. This is only possible under transdisciplinary approach where all players effectively work together for their common objective, that is, a sustainable and resilient society for human well-being.

5.6.2 Additional remarks

Integrated flood risk management (IFRM) is nothing but a process of coordinating all possible means of managing flood hazards by structural, non-structural, engineering, institutional, educational, medical, socio-economic, social capital and other related means working together. This is only possible through societal transformation where all decisions related to flood risk management are made in transparent way at all levels, from national to local communities, so that all stakeholders can participate within their committed responsibility. Obviously, it is impossible to realize such a basic policy change only in flood management and integrated water resources management (IWRM) but should extend to all societal decision-making processes as it is a societal transformation.

Living with floods

Any people in the world are living with floods, or any natural hazards in general, by their own way according to their given conditions, namely, the relation between nature and human settlements including livelihood and land use. In some regions structural means prevail where flood extremes are limited and controllable within economically and technically feasible investment. In some regions non-structural means prevail where flood extremes exceed the capacity of possible structural means within their economic and technical capacity. In general, both structural and non-structural means need to be applied in the best combination. It should be noted that while structural means can never be 100% reliable as hazards exceeding the design scale do occur and unexpected structural or human accidents are inevitable, it is impossible to rely 100% on non-structural means either. Because non-structural means such as early warning and evacuation can save life and make some business continuity possible, but cannot secure economic assets and activities, and accordingly, make accumulation of wealth impossible. By hazard maps and land use control, people can select safe place to live and for human activities only if there are ample safe places available which is not necessarily the case in most parts of the world.

Structural means include not only levees, diversion canals, storage reservoirs, retardation ponds, etc. but also basin management such as forestation, sabo works, in-situ storage and infiltration, floodproofing, etc. Non-structural means include preparedness, forecasting, warning, evacuation, rescue, life support for victims, evacuees and the affected, zoning, insurance, hazard maps, training and education (especially important where people's mobility is high), etc. For both, science and technology have an essential role to play.

Some essentials of flood management

To follow are some more aspects of flood management which have not been clearly mentioned so far or something already mentioned but in different words.

1. Flood management is part of national economy. Good management is possible under a good economy. It is a reward of peace and prosperity.

Conversely, if a nation is committed to a war or any destructive conflict, there is no relevance to make a serious effort on disaster management.

2. Flood control and damage potential are not in a vicious cycle as long as a societal system is sustainable, or as long as a society can add more safety to meet increased damage potential and is prepared for excess floods. Under a vicious cycle, provision of safety cannot catch up with increasing damage potential. But obviously this race cannot continue forever and soon or later the race should stop by stopping the increase of damage potential or taking care of the need of additional structural control of floods by additional non-structural means.

3. Both structural and non-structural means are necessary for flood risk management. Their proportion would depend basically on economy, land use and people's way of life. However, governing capacity, or institutional and administrative capacity to work together controls successful operation of both means. Especially management of non-structural means requires intensive human collaboration so that society tends to rely on large scale structural means although they also require intensive human care for their maintenance and operation. At any rate, the proportion of structural and non-structural means is not a matter of economy only but depends largely on collaboration capacity of society for both means to complement each other. Without it, neither would function enough to save lives, properties nor public services.

4. More room for water is the central option of living with floods but requires free space. The concept of compact city would help a lot into this direction supported by depopulation started now and continues to the future. Careful and endurable policy and planning towards a compact city would be necessary especially under climate, energy and environmental pressures.

5. New technology provides new options for flood management: Accurate prediction, effective preparation and warning; advanced flood-proofing for a block of area; floating houses; underground rivers; remote sensing, data transmission, cell phones, social communication medias, big data and many others. Flood and related disaster management should be sensitive and proactive to new emerging technologies.

6. There is no universal way for living with floods. Right solution would depend on natural condition, development state of society, societal preference, economic situation, etc. Societal system structure including social capital and operation of the help triangle of self-help, mutual help and the public help are different in different communities.

7. People in the Asian Pacific region share a number of common natural and social conditions such as high volcanic mountains, narrow basins, heavy rains, living on rice and high population density which are very different from continental flat land with undulating landscape and

mild rain such as in Europe. There is a regional relevance for the people in the region to exchange experiences and knowledge to mutually learn. Activities of the UNESCO IHP Regional Steering Committee (RSC) for Southeast Asia and the Pacific since 1993 have been playing a remarkable role in this purpose (UNESCO IHP RSC: website).

8. Flood insurance can serve where a large population live in flood risk areas but not hit everywhere at once but in different places and times. It is a mechanism to spread flood losses over time and space. Insurance can also serve as a flood risk reducer if the insurance premium gives an incentive for people to choose less risk areas to live or employ flood damage mitigation measures such as flood proofing. But as the damages are huge if a large area is flooded, it is necessary to have a re-insurance system supported by the Government under public-private-partnership (PPP) (Hudson et al., 2019).

9. Evolution of methodologies of Living with Floods may be summarized as follows:

1) From flood control to human adjustment.
2) From confining water into river channel and quickly draining to the sea to more space for water to retard and to infiltrate underground to slowly discharge. From riverine flood control to basin management of flood.
3) From no flooding to residential area to no human losses with early warning and evacuation.
4) From reliance on public-help to public-help supported mutual-help and self-help initiatives.
5) From a piecemeal approach to an integrated approach. A mono-disciplinary approach to a multi-disciplinary approach, an inter-disciplinary approach and to a transdisciplinary approach.

5.7 EXPERIENCES ON TSUNAMIS

Tsunami is one of the coastal floods that Japan frequently suffers together with storm surges by typhoons. It is caused mostly by earthquakes but sometimes by submarine landslides, submarine volcanic eruptions, meteorites, etc. which create a sudden movement of the bottom of the sea and travels in the speed of long wave $\sqrt{gh}$ where g is gravity acceleration 9.8 m/s² and h is depth of the sea in meter. As the average depth of the Pacific Ocean is around 4000 m, the speed is roughly 200 m/s or 720 km/h, like a jet plane and even from Chili to Japan it takes only one day. Although the tsunami velocity becomes slow near the coast, height amplifies especially when it comes into V letter-shaped bays and runs up to high slopes or cliffs.

It is not a wave but a rise of the sea as the wavelength is more than several km to several tens km. It is truly a major calamity in Japan and people have been living with it in a long history.

5.7.1 Great East Japan Earthquake (GEJE) Disaster

Outline of GEJE disaster 2011

The Great East Japan Earthquake (GEJE) Disaster that occurred on 11 March 2011 was the greatest catastrophe that Japan has ever experienced since World War II. More than 20,000 lives were lost or missing including related deaths (see Note 1) and the recovery process is still in its halfway after more than ten years. It was a cascading hazard from earthquake to tsunami and from tsunami to nuclear meltdown. It was a cascading disaster from asset to utility and from utility to livelihood damages, further, from local to nationwide, and nation to international through supply chain. Figure 5.25 illustrates such cascading effects. GEJET stands for GEJE and Tsunami as

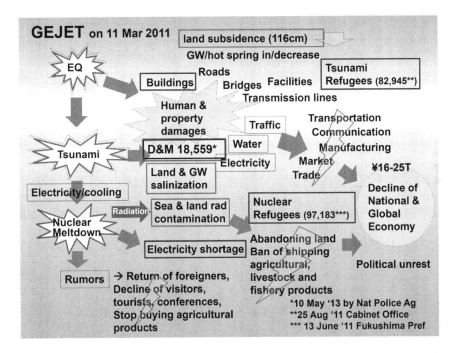

Figure 5.25 Cascade of hazards (from earthquake to tsunami and to nuclear meltdown) and cascades of disaster impacts (human and property damages to lifeline and utility, to livelihood, and to outer world through supply chain) of the Great East Japan Earthquake and Tsunami (GEJET) on 11 March 2011.

almost all losses of the GEJE disaster are due to tsunami. The total direct economic loss was estimated ¥16.9 Trillion (CO, 2016).

> Note 1: As of March 2021, the number of deaths was 15,889 and missing 2526 (NPA, 2021.3.10) and the number of related deaths was 3774 whom City, Town or Village (CTV) designated as eligible for condolence payment to their bereaved (Reconstruction Agency, 2021b, June 30). The number of deaths has been no change since 11 September 2014 while that of missing decreased by 475 since then (NPA, 2014). The related death has still been increasing.

The earthquake occurred at 14:46 on 11 March 2011 with a magnitude of 9.0 Mw at the epicenter about 130 km east-south-east offshore of Ojika Peninsula near Sendai in about 24 km depth. This caused huge tsunamis hitting all the way from Hokkaido to Kyushu inundating 561 km² areas especially in Iwate, Miyagi and Fukushima Prefectures extending to Aomori, Ibaraki and Chiba Prefectures. The tsunami height was in many areas over 7–8 m in Tohoku, e.g., in Minami Soma coast, 9 m and in Onagawa Bay, 14.5 m. The maximum run-up height observed was 40.5 m in Aneyoshi, Miyako City, Iwate Prefecture (CO: website on GEJE; Honkawa data tribune: website). The totally destroyed houses were 127,367 and half damaged 273,335 as of 11 September 2014 (NPA, 2014), and the inundated agricultural land was 236 km² (MAFF, 2014) which had salt contamination lasting for years. The number of evacuees was about 470,000 at the peak three days after the earthquake (FDMA, 2013) and still 39,515 as of 8 October 2021 (Reconstruction Agency, 2021c).

The Fukushima Daiichi Nuclear Power station is located in the coast at the border between Futaba Town in the north and Okuma Town in the south, Fukushima Prefecture. The station was hit by high tsunami resulting in the inundation height of O.P+11.5–15.5 m in the power station (see Note 2) (Government Investigation Committee, 2011), while the maximum inundation height of O.P+17 m and the maximum run-up height O.P+18 m (SCJ, 2019). Although the ground of the station itself was O.P+10 m high for no. 1–4 plants and 13 m for no. 5 and 6 plants causing no serious structural damages, the emergency sea water pumps located at O.P+4 m completely submerged and became unable to function for cooling down the plants. This caused the most serious nuclear accidents, meltdown, hydrogen explosion and radiation contamination around a large area. The total area where evacuation instruction was issued was 1150 km² of 12 cities, towns or villages (CTVs) in August 2013 (CO, 2019; Fukushima Recovery Station: website on Status of Evacuation Instruction Area) and about 165,000 persons evacuated at the peak in May 2012 and still 34,650 in October 2021 (Fukushima Prefecture, 2021; Fukushima Recovery Station: website on Status of Evacuation Instruction Area and Refugee Support).

Note 2: O.P is the reference measure at the Onahama harbor in Fukushima Prefecture of which the origin is 0.727 m below the mean sea level of Tokyo Bay.

Recovery strategy: L1 and L2 approach

The Government immediately formed the Emergency Disaster Response Headquarter headed by the Prime Minister in the Cabinet Office. Then the national government, prefectures, municipalities, news media and, above all, a huge number of individuals at each emergency site confronted respective actions of all kinds. Tsunami relief operation, nuclear meltdown control, evacuation from radiation-contaminated areas, provision of lifeline supports to emergency evacuation shelters and isolated villages, etc. All were undertaken under an uncertain reality and unforeseen future with continuing aftershocks and still increasing impacts of the disaster in a cascading manner.

While the aftermath was still continuing, the "Reconstruction Design Council in Response to the GEJE" was established in the Cabinet Office on 11 April 2011. It was an advisory panel which reported the final recommendation "Towards Reconstruction: Hope beyond the Disaster" on 25 June 2011 (Reconstruction Design Council, 2011). It was based on the seven basic principles of reconstruction, that is, the spirit of mourning, community-centered, regional latent power to lead Japan's future, disaster-resilient communities, revitalization of Japanese economy, swift resolution of nuclear accidents, and a spirit of nation's solidarity. Regarding safety of a region, it is recommended that the spatial design of a region should be made safe for the long future by the combined measures of seawalls near the coast, secondary levees inland, land use zoning for the elevated land, and migration to high land. It is important to note that there was no mention of quick recovery of the region by simply building back to the original state. The recommended strategy obviously requires a long time for planning, consensus building and construction with huge financial resources.

The institutional basis of reconstruction, the "Basic Law on Reconstruction in Response to the Great East Japan Earthquake" was issued on 24 June 2011 (Heisei 23 Law No. 76, 2011), a day before the submission of the recommendation. The law therefore well reflected the spirit of the recommendation. It declared the basic philosophy of reconstruction in Article 2 where it says:

1) . . . the reconstruction should aim to build new communities, with a vision of Japan which is appropriate for the mid-twenty-first century, based on the understanding and cooperation of the people, by promoting dramatic measures for the revitalization of a vibrant Japan which

are not limited to a recovery from the disaster by simply restoring affected facilities to their original state, as well as reconstruction measures which aim to facilitate every individual to overcome the disaster and lead a prosperous life.

About the specific measures in the same article,

5) The following measures shall be promoted:

   a) Measures to create safe communities where preventive measures against natural disaster are effective and everyone can live with a sense of security for years and decades to come.
   b) Measures to create employment opportunities in the disaster-affected regions and to restore sustainable and vibrant society and economy.
   c) Measures to promote regional culture, to maintain and strengthen bonds in communities as well as to create a society of harmonious coexistence.

6) Above measures shall be implemented for the reconstruction of areas affected by the nuclear accident, taking into account the progress made in recovery from accident.

By this Basic Reconstruction Law, the Reconstruction Headquarter was established in the Cabinet Office on 24 June 2011, which was later succeeded by the Reconstruction Agency next year on 10 February 2012. Led and coordinated by the Reconstruction Agency, the local governments, local communities and the entire Japan (including reconstruction tax) together have been implementing reconstruction on site.

The construction standard against tsunami was discussed in the Committee for Technical Investigation on Countermeasures for Earthquakes and Tsunamis Based on the Lessons Learned from the "2011 off the Pacific coast of Tohoku Earthquake" established in the Cabinet Office in May 2011 and its interim report issued on 26 June 2011 recommended a "two level approach" (Central Disaster Management Council, 2011a, 2011b). One level is an extremely low-frequency maximum scale tsunami to be managed mainly by evacuation with supporting comprehensive disaster prevention management and the other is a more frequent but large-scale tsunami to be controlled by seawalls and other coastal prevention facilities. Responding to the recommendation, MLIT named the latter Level 1 (L1) tsunami around once a hundred-year occurrence to protect all of human lives, properties and livelihood, and the former Level 2 (L2) tsunami as around once a thousand-year occurrence to protect human lives and the national economy (MLIT, 2012b).

Following this policy recommendation, "Law on Development of Areas Resilient to Tsunami Disasters" was enacted in December 2011 (Heisei 23

Law No. 123, 2011). This law ordered the MLIT Minister to make a basic guideline to make tsunami-prone areas to be protected and prefectural governors to produce tsunami hazard maps based on the L2 tsunami that identify (1) tsunami warning areas (yellow zone) where CTV governors to prepare for evacuation routes, areas and centers, and practice evacuation training, and (2) special tsunami warning areas (red zone) where development is regulated and evacuation shelters to be built.

Minamisanriku-cho, Miyagi Prefecture is one of the deadliest destroyed towns in Tohoku where the reconstruction has been taking place in three zones as schematically illustrated in Figure 5.26. Against L1 tsunami, a gigantic 8.7 m high seawall was constructed along the entire coast and rivers that protect the whole town and activities (NACSJ: website on Minamisanriku-cho). But for L2 tsunami, residential areas were either moved to high land newly opened in the nearby hillsides or to the land elevated by landfill. The coastal flat land was used only for business or recreation purposes with evacuation roads to hillsides and tall evacuation buildings. Some secondary dikes or elevated roads were also constructed.

The longest time needed for reconstruction was to reach an agreement on common matters such as how high the sea wall should be, which hillside for a community to move, which land area should be elevated for community replacement, who were the landowners, whether landowners agree etc. etc. In Japan, agreement is considered preferably consensus as much as possible which takes much time and efforts. Also, in many cases, land, especially a mountain, is owned by a number of people such as many relatives for several generations which make land registration clearance time-consuming. It is money consuming too as people have to stay in temporal house supported by tax. This example clearly demonstrates building back better is a time- and money-consuming process. This is why a pre-disaster plan is

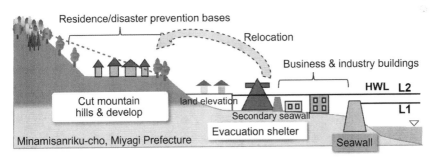

Figure 5.26 Schematic illustration of L1 and L2 approach for reconstruction of communities against tsunami. Residential areas are relocated to elevated land or higher land, and low land is used for business with various evacuation facilities and training drills. A plan of Minamisanriku-cho, Miyagi Prefecture, modified by the author. (Courtesy of Minamisanriku-cho)

necessary to shorten decision-making process and quickly start the implementation of a plan.

According to the Reconstruction Agency (2016, 2021a), at the end of March 2016, 5 years after the disaster, the completed private housing lots in the elevated land under the community relocation program were 43.4% of about 18,000 planned lots and those under the public housing program were at 57.6% of about 30,000 planned lots. It was only December 2020 when both reached 100% completion. But this was only the completed "land lots" for residence and not "houses" themselves with returned people. It takes much more time for people to come back from the evacuated places where they lived their life already for more than 10 years with jobs and schools. It is no wonder that there are many people who choose to stay in the evacuated places. The aerial photo Figure 5.27 shows that there are still many empty places left in Minamisanriku-cho.

Such speed of recovery of towns is indeed quite slow as compared with cases in other countries such as Banda Aceh, Indonesia after Indian Ocean Tsunami on 26 December 2004, and Tacloban, the Philippines after the storm surge by Typhoon Haiyan on 8 November 2013. The reason was clear from their recovery strategy that let people return to the original place to recover the original state of order as soon as possible. On the contrary in Tohoku, the main focus was to make the same disaster not happen again. To reconstruct such safe communities, the infrastructure design was set to such a high standard as stated in the law regardless of a long time necessary

Figure 5.27 Minamisanriku-cho, Miyagi Prefecture in April 2020. Nine years after the tsunami disaster, there are still many plots of empty lands waiting for proper business to come. (Photo: Courtesy of Minamisanriku-cho, NACSJ: website)

to complete. Slow recovery is a compensation to safer tomorrow and may not be possible for all nations in different economic conditions (Takeuchi and Tanaka, 2016).

5.7.2 Safety culture for tsunami

Large tsunamis in Japan

In Japan, giant tsunamis have been hitting a number of times in history. Some of the largest ones were as follows:

13 July 869 (Jogan 11): The Jogan Tsunami triggered by an estimated Mw 8.4 earthquake far offshore of Sanriku hit Taga Castle area near Sendai and killed 1000 people which was quite a large number considering low population density. A sedimentary trace by this tsunami was found at about 1.5 km inland from the coast of Namie-cho near the Fukushima Daiichi nuclear station (Namegaya et al., 2010) which after the nuclear accidents raised a regret of not having paid an attention before. Note that several years before in 864–866, the Great Jogan Eruption of Mt. Fuji occurred that released much lava and created the current form of Fuji Five Lakes by separating Seno Lake into Shoji Lake and Sai Lake (NAOJ, 2019).

28 October 1707 (Hoei 4): The Hoei Earthquake (estimated in JMA scale Mj 8.6) occurred, which was one of the biggest earthquakes Japan experienced. The epicenter extended from off-shore Enshu Sea to Shikoku Island caused by the Philippines Plate subducting from the South Sea Trough. The earthquake damages were centered in Tokaido, Ise-Bay and Kii Peninsula and huge tsunamis hit from Izu Peninsula to Kyushu Island. It is considered at least 20,000 people died (NAOJ, 2019).

16 December 1707 (Hoei 4): Only 49 days later, Mt. Fuji Hoei Eruption occurred and lasted 16 days, which produced as much as 1.7 km³ debris and ashes and resulted in long-lasting influence mainly to the northeastern Shizuoka to northwestern Kanagawa Prefectures including the Sakawa River which caused floods due to riverbed rise as will be mentioned in Epilogue. It was the last eruption of Mt. Fuji till now. Although the relation between earthquakes and volcanic eruptions is not necessarily related in general, this was a rare case that it is natural to consider induced by the Hoei Earthquake (Koyama, 2006; Matsuura, 2014).

24 April 1771 (Meiwa 8): The Great Meiwa Tsunami hit Ishigaki and Yaeyama Islands, Okinawa which took about 12,000 lives and the impacts of disease and famine due to salinization lasted a hundred years (JMA Okinawa: website on the Great Meiwa Tsunami).

23 December 1854 (Ansei 1): The Ansei East Sea Earthquake occurred and caused a devastating tsunami which hit from the Bousou Peninsula to Shikoku Island and took 2–3 thousand lives (CO, 2005a; National Astronomical

Observatory, 2021). A Russian frigate "Diana" was broken in the Shimoda harbor and sank offshore of the Fuji River estuary, Suruga Bay.

24 December 1854 (Ansei 1): Only 31 hours later, the Ansei South Sea Earthquake occurred and caused another devastating tsunami which hit from Tokai to Kyushu coasts and killed some thousand people. In Hiro Village, now in Wakayama Prefecture, HAMAGUCHI Goryo saved villagers by firing the just-harvested rice pile to guide them on where to evacuate. His story will be introduced in the next section.

15 June 1896 (Meiji 29): The Meiji Sanriku Tsunami triggered by Offshore of Sanriku Earthquake of Mw 8.2 hit the coasts from Hokkaido to Ojika Peninsula and the dead or missing were 21,959 people. In Taro Village (now in Miyako City, Iwate Prefecture), all 345 houses were fully or half destroyed and 1867 persons (83.1%) of villagers were dead or missing. In Kamaishi Town and Toni Village (now both in Kamaishi City, Iwate Prefecture), 3765 (53.9%) and 1684 (66.4%) of the population, respectively, were dead or missing. Shizukawa Town, now Minamisanriku-cho (Town), Miyagi Prefecture that was mentioned in the previous section lost 371 (20.9%) persons (Yamashita, 2008).

3 March 1933 (Showa 8): The Showa Sanriku Tsunami triggered by another Offshore of Sanriku Earthquake of Mw 8.1 hit the Pacific coast especially Sanriku again and the dead or missing were 3064. The magnitude of the tsunami was considered roughly about 75% of the Meiji Sanriku Tsunami as the highest run-up in Shirahama in Ryori Village (now in Ofunato City, Iwate Prefecture) was 28.7 m as compared with 38.2 m at the same place in the Meiji Sanriku Tsunami. The number of houses fully or half destroyed by tsunami was also in the same proportion, i.e., about 6000 in Showa as compared with 8000 in the Meiji. Then what about the dead and missing? In Taro Village, it was 972 (20%) persons of the population, in Kamaishi Town and Tanni Village, 38 (0.15%) and 359 (9.7%) and in Shizukawa Town, it was 0 (Yamashita, 2008). Since it was only 37 years later, the previous memory was still alive and communities were better prepared to response quick that made the damage much smaller.

There were many more tsunamis that resulted in tragic consequences to coastal communities. Some left important lessons to future generations and served for building tsunami culture. Among them two cases, Aneyoshi's memorial stone and Inamura no Hi are introduced in the following.

Great Tsunami memorial stone in Aneyoshi: Never build a house below here

In Aneyoshi, Omoe Village (now in Miyako City, Iwate Prefecture), the 1896 Meiji Sanriku Tsunami height was 18.9 m and all of 11 houses were fully destroyed losing all 78 population but two. Nevertheless, people such as their relatives living in other places came back and rebuilt houses in the same place and 37 years later, received the 1933 Showa Sanriku Tsunami of 12.4 m high. Again, they lost all 14 houses and 111 population except four who were luckily out of village. Among the dead, included were about 50 people who were working in the beach on fixed shore net fishing (Iwate Nippo, 2018). This repeated tragedy made them to build a memory stone nearby 60 m high place as seen in Figure 5.28. It reads:

Great Tsunami Memorial Monument

- Houses on high ground are the peace and prosperity of future generations.
- Remember the huge tsunamis that gave us tragic damages.
- Never build a house below here.

Tsunami came up here both Meiji 29 (1896) and Showa 8 (1933) and the village was totally destroyed leaving the alive only 2 and 4, respectively. Be cautious forever.

The stone was their strong wish for future generations but at the same time was an expression of great regret for their mistake having repeated the same tragedy again. This time relatives returned built houses in a high place about 500 m from the beach. Because of this in the Great

Figure 5.28 Great tsunami memorial stone saying "Never build a house below here" built after the Showa Sanriku Tsunami 1933 at Aneyoshi, Miyako City, Iwate Prefecture. (Photo taken by TANAKA Shigenobu on 27 September 2007)

East Japan Earthquake, although the tsunami height was 38 m and the run-up came up to about 50 m to the stone in the coast side, nobody died (Iwate Nippo, 2018). Following previous lessons was shown so important. In fact, such stones were built about 200 in Sanriku after the Showa Sanriku Tsunami using part of voluntary donations they received (Yamashita, 2008).

"Inamura no Hi"

It was around 10 am on 23 December 1854 when a big shock (the Ansei East Sea Earthquake) was felt and HAMAGUCHI Goryo immediately evacuated his family and villagers to the Hiro Hachiman Shrine in the back hill of his house in Hiro Village of Kishu Domain (now in Hirogawa Town, Wakayama Prefecture) but no tsunami came. Next day after 4 pm on 24 December, another shock (the Ansei South Sea Earthquake) incomparably stronger than the previous day came and soon after the cease of tremble, a great tsunami, over 5 m hit the Hiro Harbor destroying many boats and breaking down fish factories and houses near the beach, upstreaming the Hiro River, flooded streets and fields. Goryo himself could barely manage to run up to a safe place but realized his fellow villagers were drowning and roads and fields were washed away by repeated waves which made villages unable to identify where they were and the way to evacuate to the higher place. As the sunset time on 24 December in Wakayama is around 16:55 so that soon after 17:00 it must have been complete dark. The Goryo put *hi* (fire) on *inamura* (a pile of harvested rice for drying in rice fields) and made villagers to see the way to evacuate to the Hiro Hachiman Shrine. By his wise, timely and brave action, it was considered that he saved over 300 villagers (Thompson, 2020; Hirogawa Town: website on Inamura no Hi Museum).

After this tsunami devastation, he led the construction of 5-m-high and 600-m-long sea dikes along the coast and planted pine trees in the seaside. Although the work was supported by the public budget, he also contributed his private money as much as 94 kan (≈350 kg) silver. It took 4 years with 56,736 person-day to complete which supported livelihood of villagers after losing everything in devastation (CO, 2005b; Disaster Prevention System Institute: website on Inamura no Hi). Ninety-two years later, this dike saved villagers from 4–5 m tsunami of Showa South Sea Earthquake on 21 December 1946.

HAMAGUCHI Goryo was born in 1820 in Hiro Village in a branch family of the Yamasa Soy Sauce owners and became the 7th head of the main family to run the Yamasa Company in Choshi, Soushu (now in Chiba Prefecture). At the time of the Ansei East Sea and South Sea Earthquakes in 1854, he was 35 years old (counting the birth year as one) and temporarily back in Hiro Village, Kishu. So he was a leader of the village. Later, he was appointed as the treasurer of Kishu Domain and in the Meiji Era elected as the first chairman of the Wakayama Prefectural Parliament. He died in New York during his trip to the USA.

The story of Goryo's wise and selfless action to save fellow villagers at the tsunami emergency of the Ansei South Sea Earthquake greatly impressed a US writer Lafcadio Hearn who came to Japan about 35 years later in 1890. Only 5 years after Lafcadio arrived in Japan, another truly devastating Meiji Sanriku Tsunami occurred in June 1896. When he heard of its tragic consequences, it stimulated him to write an essay about Goryo's wise action at the Ansei South Sea Tsunami under the title "A Living God". He put this essay as an opening story of his book "Gleanings in Buddha-Fields" published in Boston and London in September 1896, only three months later from the Meiji Sanriku Tsunami (Inamura no Hi: website on A Living God).

Nevertheless, it took 40 years until this essay became known to the Japanese. An elementary school teacher NAKAI Tsunezo who was in the town neighboring to Hiro Town where Goryo's house was, learned "A living god" in Taikyu Middle High School built by Goryo and got greatly moved. When the Ministry of Education offered a competition for an essay to be put into an elementary school textbook in 1934, a year after the Showa Sanriku Tsunami, he wrote a Goryo's story inspired by "A Living God" under the title "Inamura no Hi (fire on the pile of just harvested rice)" and applied. He won the competition and the essay was put to the nation's elementary school textbook of the 5th grade from 1937 to 1947. Although the story became a bit different from the reality by the two authors' respective modifications, the importance of evacuation at tsunami, leadership for evacuation and community preparedness was well delivered to many pupils.

Thus "Inamura no Hi" now became a symbolic teaching material of the needs of evacuation and community preparedness at tsunami in Japan as well as in the world. According to the morning print of Mainichi Newspaper in Osaka on 26 June 1994, in the 4th-grade textbook "Spinners" in an elementary school in the suburbs of Colorado Spring, US, there was an article of "Inamura no Hi" under the title "The burning of the Rice Fields" (Inamura no Hi: website on Foreign Material).

Talking a bit more about the key person Lafcadio Hearn, he came to Japan with his deep interest in Japanese culture but what made him stay in Japan seems to have an accidental fortune behind. Lafcadio was born in Greece in 1850, brought up in Ireland and migrated to the USA in 1869 where he eventually became successful as a writer of newspaper "Times-Democrat" in New Orleans. On the occasion of the World's Industrial and Cotton Centennial Exposition in New Orleans in 1884–1885, he was much interested in the expositions of Japan Pavilion and there he met HATTORI

Ichizo who was the representative of the Japanese Government as well as an earthquake and tsunami scientist served as the first president of Seismological Society of Japan from 1880 to 1882 (Nakagawa, 2013).

Lafcadio came to Japan to gather news for the newspaper in 1890 but resigned the company due to some dis-satisfaction. Then Hattori, who was one of the bureau directors of the Ministry of Education at that time, arranged him an English teacher's position in Shimane Middle High School and Teacher's College. The Meiji Sanriku Earthquake and Tsunami disasters happened on 15 June 1896 and "A living God" was published only three months later in September. At this Tsunami hit, Hattori was the governor of Iwate Prefecture taking the frontline responsibility on emergency response. Unfortunately, there is no document found to indicate whether Lafcadio had any communication with him or not on this tsunami disaster.

Lafcadio became quite famous and respected in Japan as a writer. He introduced Japanese culture to the world through his books published under his Japanese name KOIZUMI Yakumo. The most famous one was "Kwaidan", a collection of old Japanese horror stories published in 1904, the year of his decease at the age of 54 (Lafcadio Hearn Memorial Museum: website on KOIZUMI Yakumo).

REFERENCES

Abe, Akihiro (2014) *Study on "Boys Be Ambitious"*. http://culture-h.jp/hatadake-katsuyo/bun69BoysBeAmbitious.pdf (accessed 22 February 2021).

Ando, Yoshihisa, Y. Yasoshima and K. Musiake (1976) *Flood Control and Land Use in New Towns*. Proc. JSCE Annual Convention. http://library.jsce.or.jp/jsce/open/00035/1976/31-04-0220.pdf (accessed 9 December 2020).

Archives Tone River Editorial Committee (2001, November 30) *Archives Tone River*. Ed. supervision by Tadashi Miyamura. Shinzansha-Scitech, Tokyo.

Beck, H.E., N.E. Zimmermann, T.R. McVicar, N. Vergopolan, A. Berg and E.F. Wood (2018) Present and Future Köppen-Geiger Climate Classification Maps at 1-km Resolution. *Scientific Data*, 5, Article number: 180214. http://doi.org/10.1038/sdata.2018.214.

CDMC (2007) *1944 Southeastern Sea Earthquake and 1945 Mikawa Earthquake*. Special Investigation Committee on Inheriting Disaster Lessons, Central Disaster Management Council, Cabinet Office, March 2007. www.bousai.go.jp/kyoiku/kyokun/kyoukunnokeishou/rep/1944_tounankai_jishin/index.html (accessed 5 December 2021).

CDMC (2008) *1959 Ise-Bay Typhoon Report*. Special Investigation Committee on Inheriting Disaster Lessons, Central Disaster Management Council, Cabinet Office, March 2008. www.bousai.go.jp/kyoiku/kyokun/kyoukunnokeishou/rep/1959_isewan_typhoon/pdf/07_chap3-1.pdf (accessed 10 February 2021).

CDMC (2010, January) *1947 Kathleen Typhoon report. Special Investigation Committee on Inheriting Disaster Lessons*. Central Disaster Management Council, Cabinet Office. www.bousai.go.jp/kyoiku/kyokun/kyoukunnokeishou/rep/1947_kathleen_typhoon/index.html (accessed 31 May 2021).

Central Disaster Management Council (2011a) *An Interim Report of the Committee for Technical Investigation on Countermeasures for Earthquakes and Tsunamis Based on the Lessons Learned from the "2011 off the Pacific coast of Tohoku Earthquake"*, 26 June 2021. www.bousai.go.jp/kaigirep/chousakai/tohokukyokun/pdf/tyuukan.pdf (accessed 6 November 2021).

Central Disaster Management Council (2011b) *Report of the Committee for Technical Investigation on Countermeasures for Earthquakes and Tsunamis Based on the Lessons Learned from the "2011 off the Pacific Coast of Tohoku Earthquake"*, 28 September 2011. www.bousai.go.jp/kaigirep/chousakai/tohokukyokun/pdf/Report.pdf (accessed 31 May 2019).

Chiba Prefecture: website on Super Levee at Yako. www.pref.chiba.lg.jp/kendosei/stock/h29/documents/11_1superteibou_h29.pdf (accessed 20 January 2022).

CO (2005a) *1854 Ansei East Sea Earthquake and Ansei South Sea Earthquake*. A Report of Special Committee on Inheriting Disaster Lessons. www.bousai.go.jp/kyoiku/kyokun/kyoukunnokeishou/rep/1854_ansei_toukai_nankai_jishin/index.html (accessed 13 November 2021).

CO (2005b) *Tsunami and Inamura no Hi*. *Kouhou "Bousai", no. 26, 2005/3*, 14–15. www.bousai.go.jp/kyoiku/kyokun/kyoukunnokeishou/pdf/kouhou026_14-15.pdf (accessed 13 November 2021).

CO (2016) *Disaster Prevention White Paper, 2016*. Table 19. www.bousai.go.jp/kaigirep/hakusho/h28/honbun/3b_6s_19_00.html (accessed 5 November 2021).

CO (2019) *On Status of Areas Under Evacuation Instruction*. Life Support Team for Nuclear Power Accidents' Refugees, Cabinet Office, July 2019. www.mext.go.jp/content/1421518_03.pdf (accessed 6 November 2021).

CO: website on GEJE. www.bousai.go.jp/kohou/kouhoubousai/h23/63/special_01.html (accessed 2 November 2021).

Disaster Portal: website on Hazard Maps to Be Overlaid. https://disaportal.gsi.go.jp/maps/?ll=35.661759,138.570213&z=11&base=pale&vs=c1j0l0u0 (accessed 9 June 2021).

Disaster Prevention System Institute: website on Inamura no Hi. www.bo-sai.co.jp/inamuranohi.htm (accessed 13 November 2021).

Edo River Office: website on the Metropolitan Area Outer Underground Discharge Channel. *Kanto Development Bureau, MLIT*. www.ktr.mlit.go.jp/edogawa/edogawa00402.html (accessed 4 August 2020).

Edo River Office: website on the Work of the Office. *Kanto Development Bureau, MLIT*. www.ktr.mlit.go.jp/edogawa/edogawa00402.html (accessed 4 August 2020).

Endo, T., S. Kawashima and M. Kawai (2001) Historical Review of Development of Land Subsidence and Its Cease in Shitamachi Lowland, Tokyo. *Journal of the Japan Society of Engineering Geology*, 42(2), 74–87. www.jstage.jst.go.jp/article/jjseg1960/42/2/42_2_74/_pdf (accessed 17 February 2021).

Environmental Agency (2000) *Environmental White Paper, 2000*. Figure 1–3–1. www.env.go.jp/policy/hakusyo/img/212/fb2.1.3.1.gif (accessed 22 February 2022).

e-Stat: website on Crop Production. Statistics Bureau of Japan, Ministry of Internal Affairs and Communications. www.e-stat.go.jp/stat-search/files?page=1&layout=datalist&toukei=00500215&tstat=000001013427&tclass1=000001032288&tclass2=000001034728 (accessed 5 June 2021).

e-Stat: website on Water-related Disaster Damages. Statistics Bureau of Japan, Ministry of Internal Affairs and Communications. www.e-stat.go.jp/dbview?sid=0003161327 (accessed 2 June 2021).

FDMA (2013) *GEJE Disaster Records. Fire and Disaster Management Agency*, March 2013. Chapter 3.6. www.fdma.go.jp/disaster/higashinihon/item/higash-inihon001_21_03-06.pdf (accessed 6 November 2021).

FDMA (2019, January 1) *Damages and Responses Situations of 2018 July Heavy Rains and Typhoon 12.* (Report no. 59th) At 13:00 on 9 January 2018, Japan Fire and Disaster Management Agency (FDMA), Emergency Response Office. h30–7_59.pdf (fdma.go.jp) (accessed 21 December 2020).

FRICS (1997) *Flood Inundation Risk Area Map. 3.* The Upper Tone River and the Edo River. pp. 9–10. Supervised by Ministry of Construction, April 1997.

FRICS: website on Radar Rain Gauges. www.river.or.jp/post_22.html (accessed 30 July 2020).

Fujibe, Fumiaki (2018) Strong Winds and Disaster in Eastern Japan Due to the Typhoon on 28 September 1902. *Tenki*, 65(10), 29–37. www.metsoc.jp/tenki/pdf/2018/2018_10_0029.pdf (accessed 5 December 2021).

Fujinawa, K. (1988, March) Groundwater Use for Melting Snow in the Nagaoka Plain. *Urban Kubota*, 27, 52–57. www.kubota.co.jp/siryou/pr/urban/pdf/27/index.html (accessed 18 February 2021).

Fukushima Prefecture (2021) *2011 GEJE Disaster Damage Quick Report, 1782*, 5 November 2021. www.pref.fukushima.lg.jp/uploaded/attachment/483951.pdf (accessed 6 November 2021).

Fukushima Recovery Station: website on Status of Evacuation Instruction Area. www.pref.fukushima.lg.jp/site/portal/list271-840.html (accessed 6 November 2021).

Fukushima Recovery Station: website on Status of Evacuation Instruction Area and Refugee Support. www.pref.fukushima.lg.jp/site/portal/list271.html (accessed 6 November 2021).

Gifu Prefecture: website on Heavy Rain Disaster on 12 September 1976. www.pref.gifu.lg.jp/page/6965.html (accessed 8 December 2020).

Government Investigation Committee (2011) *The Interim Report.* Investigation Committee on the Accident at the Fukushima Nuclear Power Stations, 26 December 2011. www.cas.go.jp/jp/seisaku/icanps/eng/index.html (accessed 6 November 2021).

Heisei 9 Law No. 69 (1997) *Amendment to Showa 39 River Law.* www.shugiin.go.jp/internet/itdb_housei.nsf/html/houritsu/14019970604069.htm (accessed 23 February 2022).

Heisei 9 Law No. 81 (1997, December 25) *Environmental Impact Assessment Law.* https://elaws.e-gov.go.jp/document?lawid=409AC0000000081 (accessed 19 February 2022).

Heisei 12 Law No. 57 (2000, May 8) *Sedimentary Disaster Prevention Law.* https://elaws.e-gov.go.jp/document?lawid=412AC0000000057_20210715_503AC0000000031 (accessed 19 February 2022).

Heisei 12 Law No. 87 (2000, May 19) *Special Law on Public Use of Deep Underground.* https://elaws.e-gov.go.jp/document?lawid=337AC0000000100_20150801_000000000000000 (accessed 19 February 2022).

Heisei 13 Law No. 46 (2001) *Amendment to Flood Fighting Law of 1949.* https://hourei.ndl.go.jp/simple/detail?lawId=0000090330&current=-1#text_of_the_law (accessed 10 March 2022).

Heisei 23 Law 76 (2011) *The Basic Law on Reconstruction in Response to the Great East Japan Earthquake 2011.* www.japaneselawtranslation.go.jp/law/detail/?vm=04&re=01&id=2434 (accessed 17 November 2021).

Heisei 23 Law No. 123 (2011) *Law on Development of Areas Resilient to Tsunami Disasters.* https://elaws.e-gov.go.jp/document?lawid=423AC0000000123 (accessed 31 January 2022).

Hirogawa Town: website on Inamura no Hi Museum. www.town.hirogawa.wakayama.jp/inamuranohi/siryo_goryo.html (accessed 13 November 2021).

Hiroshima for Peace: website on Kyoto University Research Team Memorial. https://hiroshimaforpeace.com/kyoto-university-research-group/ (accessed 31 May 2021).

Hiroshima Prefecture: website on A-bomb Damages. www.city.hiroshima.lg.jp/site/atomicbomb-peace/9399.html (accessed 31 May 2021).

Hiroshima Prefecture: website on Makurazaki Typhoon. www.pref.hiroshima.lg.jp/soshiki/100/makurazaki.html (accessed 31 May 2021).

Honkawa Data Tribune: website on Great East Japan Earthquake Tsunami Height. http://honkawa2.sakura.ne.jp/4363b.html (accessed 2 November 2021).

Hudson, Paul, W.J. Wouter Botzen and Jeroen C.J.H. Aerts (2019) Flood Insurance Arrangements in the European Union for Future Flood Risk Under Climate And Socioeconomic Change. *Global Environmental Change*, 58, 101966, September 2019. https://doi.org/10.1016/j.gloenvcha.2019.101966

Inamura no Hi: website on Foreign Material. http://inamuranohi.jp/library/%e6%af%8e%e6%97%a5%e6%96%b0%e8%81%9e%e3%80%80%e5%a4%a7%e9%98%aa%e6%9c%9d%e5%88%8a%e3%80%801994%e5%b9%b46%e6%9c%8826%e6%97%a5%e4%bb%98%e3%80%80%e3%80%8c%e3%83%8f%e3%83%bc%e3%83%b3%e3%81%8c%e8%a6%8b/#more-2024 (accessed 29 December 2021).

Inamura no Hi: website on A Living God. http://inamuranohi.jp/a-living-god/a-living-god-text/ (accessed 29 December 2021).

Iwate Nippo (2018) Memory of Stone Monument—Heavy Warning for Community Survival. *Iwate Daily Newspaper*, 16 October 2018. www.iwate-np.co.jp/content/ishibumi/20181016/ (accessed 13 November 2021).

Izumi, Kiyoshi (1990) Flood Control in Metropolitan Tokyo by Underground Rivers. *JJSHWR*, 3(2), 46–47. https://doi.org/10.3178/jjshwr.3.2_46 (accessed 8 August 2020).

Japan Bosai Platform: website on Implant Method. www.bosai-jp.org/ja/solution/detail/1/category (accessed 20 January 2022).

JMA: website on Correspondence Between the Warning Level and the Warning Information. www.jma.go.jp/jma/kishou/know/bosai/alertlevel.html (accessed 14 August 2020).

JMA: website on Historical Ranking of Extreme Records. www.data.jma.go.jp/obd/stats/etrn/view/rankall.php (accessed 29 September 2020).

JMA: website on Kathleen Typhoon. www.data.jma.go.jp/obd/stats/data/bosai/report/1947/19470914/19470914.html (accessed 31 May 2021).

JMA: website on List of JMA Named Natural Hazards. www.jma.go.jp/jma/kishou/know/meishou/meishou_ichiran.html (accessed 29 May 2021).

JMA: website on Makurazaki Typhoon. www.data.jma.go.jp/obd/stats/data/bosai/report/1945/19450917/19450917.html (accessed 31 May 2021).

JMA: website on Typhoon Della. www.data.jma.go.jp/obd/stats/data/bosai/report/1949/19490618/19490618.html (accessed 5 June 2021).

JMA: website on Typhoon Ion. www.data.jma.go.jp/obd/stats/data/bosai/report/1948/19480915/19480915.html (accessed 5 June 2021).

JMA: website on Typhoon Jane. www.data.jma.go.jp/obd/stats/data/bosai/report/ 1950/19500903/19500903.html (accessed 6 June 2021).

JMA: website on Typhoon Kitty. www.data.jma.go.jp/obd/stats/data/bosai/report/ 1949/19490831/19490831.html (accessed 6 June 2021).

JMA: website on Typhoon Ruth. www.data.jma.go.jp/obd/stats/data/bosai/report/ 1951/19511010/19511010.html (accessed 7 June 2021).

JMA: website on Typhoons. www.jma.go.jp/jma/kishou/know/typhoon/1-1.html (accessed 1 December 2020).

JMA Okinawa: website on the Great Meiwa Tsunami. www.jma-net.go.jp/ishigaki/ know/jishin/meiwa.html (accessed 13 November 2021).

JSCE (2012) *Knowledge on Rivers and River Engineering in Japan: The Tone River.* Ed. Supervision by Tadashi Yamada. Hydraulic Engineering Committee, JSCE.

JWA: website on the Tone River Canal. www.water.go.jp/kanto/tone/about/history/ index.html (accessed 5 December 2020).

Kajiwara, Kenji (2021) *Japanese Modern River Administration—Evolution of Policies and Laws: 1868–2019.* Horitsu Bunkasha, Kyoto.

Kanamori, W., Y. Hurusawa, M. Kimura and K. Nisizono (1995) *Sayama Dam: History of Civil-Engineering from Ancient Bank.* JSCE. www.jstage.jst.go.jp/ article/journalhs1990/15/0/15_0_483/_pdf/-char/ja (accessed 23 January 2022).

Kanto Regional Development Bureau: Pamphlet on Tone River. MLIT, Published after 2000.

Kanto Regional Development Bureau: website on Kathleen Typhoon Special. MLIT. www.ktr.mlit.go.jp/river/shihon/index00000055.html (accessed 31 May 2021).

Katada, Toshitaka (1999) *Evacuation Behavior and the Use of Hazard Maps at the Time of Abukuma River Flood of the End of August 1998.* Textbook of Disaster Management Seminar '99. Organized by Tohoku Regional Development Bureau, Ministry of Construction, pp. 17–28.

Kiguchi, M. and T. Oki (2010) Point Precipitation Observation Extremes in the World and Japan. *JJSHWR*, 23(3), 231–247.

Kinki Regional Development Bureau: website on River Management Basic Policy and River Management Implementation Plan. MLIT. www.kkr.mlit.go.jp/ wakayama/ryuiki_iinkai/ryuiki/study/pdf/2.pdf (accessed 19 January 2022).

Kofu River Road Office: website on History of the Midai River. www.ktr.mlit.go.jp/ koufu/koufu00141.html (accessed 18 September 2021).

Kofu River Road Office: website on Traditional River Works, Kanto Regional Bureau, MLIT. www.ktr.mlit.go.jp/koufu/koufu00145.html (accessed 3 December 2020).

Komae City (1975) *The Tama River Bank Breach Report.* Dai Nippon Printing Co., Tokyo.

Komae City: website on the Nightmare of the Tama River Bank Breach. www.city. komae.tokyo.jp/index.cfm/45,336,349,2102,html (accessed 8 December 2020). Also see Rakuten blog: website on the Tama River Flood, 1974. https://plaza. rakuten.co.jp/chizulove/diary/201409250000/ uploaded on 25 September 2014 (accessed 8 December 2020).

Kotobank: website on Typhoon Kitty. https://kotobank.jp/word/%E3%82%AD% E3%83%86%E3%82%A3%E5%8F%B0%E9%A2%A8-828558 (accessed 6 June 2021).

Koyama, Masato (2006) Symptom of the Hoei Eruption. Section 1.3, Chapter 2. Process and Products of the Hoei Eruption. In: Central Disaster Management Council

(2006) *Report on 1707 Mt. Fuji Hoei Eruption. Special Investigation Committee on Inheriting Disaster Lessons*, March 2006. www.bousai.go.jp/kyoiku/kyokun/kyoukunnokeishou/rep/1707_houei_fujisan_funka/pdf/1707-houei-fujisanFUN1KA_06_chap2.pdf in www.bousai.go.jp/kyoiku/kyokun/kyoukunnokeishou/rep/1707_houei_fujisan_funka/index.html (accessed 3 January 2022).

Koyama, Takaaki and Shoichi Fujita (1984) *Plan and Design of New-type Sewage System*. Kajima Institute Publishing Co., Tokyo.

Kumai, Hisao (1994, July) Volcanic Activities in the Middle Pleistocene and Transition of the Lake and the Basin. *Urban Kubota*, 33, 14–28. www.kubota.co.jp/siryou/pr/urban/pdf/33/index.html (accessed 3 December 2020).

Lafcadio Hearn Memorial Museum: website on Life of KOIZUMI Yakumo. www.hearn-museum-matsue.jp/hearn.html (accessed 13 November 2021).

MAFF (2014) *White Paper on Foods, Agriculture and Rural Villages, FY2012*. www.maff.go.jp/j/wpaper/w_maff/h25/h25_h/trend/part1/chap4/c4_1_01.html (accessed 5 November 2021).

Man-no Town Education Committee (2008) *Man-no Pond Comprehensive Investigation Report*. http://doi.org/10.24484/sitereports.72557 (accessed 23 January 2022).

Matsuura, Ritsuko (2014) Outlook of the Hoei Earthquake. Chapter 1 In: Cabinet Office Disaster-in-charge (2014) *Report on 1707 Houei Earthquake. Investigation Committee on Disaster Lessons on the Hoei Earthquake*, March 2014. www.bousai.go.jp/kyoiku/kyokun/kyoukunnokeishou/rep/1707_houeijishin/pdf/05_chap01.pdf in www.bousai.go.jp/kyoiku/kyokun/kyoukunnokeishou/rep/1707_houeijishin/ (accessed 3 January 2022).

Matsuura, S. (2012) Transition of the Post-war Tone River Flood Control Plan. *Water Science*, 324, 107–145. 08460102.mcd (jst.go.jp) (accessed 31 May 2021).

Meiji 29 Law No. 71 (1896) *River Law*. www.digital.archives.go.jp/img/153037 (accessed 23 February 2022).

Meiji 30 Law No. 29 (1897) *Sabo Law*. www.digital.archives.go.jp/img/155144 (accessed 23 February 2022).

Meiji 30 Law No. 46 (1897) *Forest Law*. www.digital.archives.go.jp/img/155278 (accessed 23 February 2022).

Metropolitan Tokyo (2016) *The Sumida River Basin of the Ara River System Management Plan*. www.kensetsu.metro.tokyo.lg.jp/temporary/content3/000007325.pdf (accessed 20 January 2022).

Metropolitan Tokyo: website on Flood Control Works of Small to Medium Scale Rivers. www.kensetsu.metro.tokyo.lg.jp/jigyo/river/chusho_seibi/index.html (accessed 12 December 2020).

Metropolitan Tokyo: website on the Underground Rivers Under the Kanda River and the 7th Circular Road. www.kensetsu.metro.tokyo.lg.jp/content/000046116.pdf (accessed 8 August 2020).

Ministry of Environment (1971) *White Paper on Environment, 1970*. www.env.go.jp/policy/hakusyo/s46/index.html (accessed 6 June 2021).

Miyoshi, Norimasa (2015) Study on National Compensation Responsibility to Flood Disaster and Basin Flood Control. *Yamanashi Gakuin University Law Journal*, 10, 115–163. Departmental Bulletin Paper. https://ygu.repo.nii.ac.jp/?action=pages_view_main&active_action=repository_view_main_item_detail&item_id=3248&item_no=1&page_id=4&block_id=82 (accessed 22 December 2021).

MLIT (2002) *Outline of River and Basin of the Fuji River System.* River Bureau of MLIT, November 2002. www.mlit.go.jp/river/shinngikai_blog/shaseishin/kasenbunkakai/shouiinkai/kihonhoushin/021115/pdf/s2_2.pdf (accessed 18 September 2021).

MLIT (2003a) *Assessment Report on Programs of Dam Projects.* MLIT River Bureau, January 2003. www.mlit.go.jp/river/dam/main/shinngikai/kondankai/dam/pubcom/pubcomt.html (accessed 12 February 2021).

MLIT (2003b) 2. Current Status of Comprehensive Flood Control Measures. *Investigation Materials for Program Assessment on Comprehensive Flood Control Measures.* 1st Meeting Material on 28 August 2003. www.mlit.go.jp/river/shinngikai_blog/past_shinngikai/gaiyou/seisaku/sougouchisui/pdf/2_1haikei_keii.pdf (accessed 12 June 2021).

MLIT (2003c) 2.2 Framework, Current Status and Effect of Comprehensive Flood Management Measures. *Investigation Materials for Program Assessment on Comprehensive Flood Control Measures.* 1st Meeting Material on 28 August 2003. www.mlit.go.jp/river/shinngikai_blog/past_shinngikai/gaiyou/seisaku/sougouchisui/pdf/2_2genjou.pdf (accessed 12 June 2021).

MLIT (2004) *Program Assessment on Comprehensive Flood Control Measures by Integrating River Basin.* March 2004. www.mlit.go.jp/river/shinngikai_blog/past_shinngikai/gaiyou/seisaku/sougouchisui/pdf/hyokasho_final.pdf (accessed 12 June 2021).

MLIT (2005a) *Water Resources in Japan, 2004.* www.mlit.go.jp/tochimizushigen/mizsei/hakusyo/h16/2-1.pdf (accessed 1 December 2020).

MLIT (2005b, October 13) *Current Status of Storm Surge Prevention in the Zero-meter Zone in Japan.* www.mlit.go.jp/river/shinngikai_blog/past_shinngikai/shinngikai/takashio/index.html (accessed 13 February 2021).

MLIT (2005c) *Flood Hazard Maps Producing Manual.* Water and Disaster Management Bureau, MLIT, June 2005. www.pwri.go.jp/icharm/publication/pdf/2005/flood_hazard_mapping_manual_jp.pdf (accessed 10 March 2022).

MLIT (2006) *Water Resources in Japan, 2006.* Division of Water Resources, MLIT, p. 217. www.mlit.go.jp/tochimizushigen/mizsei/hakusyo/H18/sankou05.pdf (accessed 8 June 2021).

MLIT (2012a, February 8) *The Post Audit of the Naka River and the Ayase River Basin Work (the Metropolitan Area Outer Underground Discharge Channel).* www.ktr.mlit.go.jp/ktr_content/content/000053312.pdf (accessed 4 August 2020).

MLIT (2012b) *MLIT White Paper, Section 3 Policy Change after the Earthquake Disaster.* www.mlit.go.jp/hakusyo/mlit/h23/hakusho/h24/html/n1131000.html (accessed 10 November 2021).

MLIT (2014, February) *Water Disasters Statistics, 2011.* Water and Disaster Management Bureau, MLIT. https://dl.ndl.go.jp/info:ndljp/pid/9667051 (accessed 5 June 2022). E-data available at e-Sat website for casualties, https://www.e-stat.go.jp/dbview?sid=0003161327 and for economic losses, https://www.e-stat.go.jp/dbview?sid=0003153065

MLIT (2015) *Design Method for Anticipated Maximum Scale Precipitation to Produce Anticipated Inundation (Floods and Inland Floods) etc,* Water and Disaster Management Bureau, MLIT, July 2015. https://www.mlit.go.jp/river/shishin_guideline/pdf/shinsuisoutei_honnbun_1507.pdf (accessed 14 April 2019).

MLIT (2018) *Engineering Guideline and Explanation of the Use of Deep Underground*. Urban Policy Division, MLIT Urban Bureau, March 2018.

MLIT (2019a) *Current Water Resources in Japan, 2019*. www.mlit.go.jp/mizukokudo/mizsei/mizukokudo_mizsei_tk2_000027.html (accessed 1 December 2020).

MLIT (2019b) *On Use of Dams*. Water and Disaster Management Bureau, 26 November 2019. www.kantei.go.jp/jp/singi/kisondam_kouzuichousetsu/dai1/siryou1.pdf (accessed 21 February 2022).

MLIT (2021a) *Current Water Resources in Japan, 2020*. Division of Water Resources, MLIT. www.mlit.go.jp/common/001442662.pdf (accessed 29 October 2021).

MLIT (2021b) *Water Disasters Statistics, 2019*. Water and Disaster Management Bureau, MLIT.

MLIT (2021c) *Status of Sedimentary Disaster Risk Area Designation in the Nation*. www.mlit.go.jp/mizukokudo/sabo/content/001444313.pdf (accessed 20 February 2022).

MLIT (2022) *Current Water Resources in Japan 2021*. Chapter 2. Water and Disaster Management Bureau. www.mlit.go.jp/mizukokudo/mizsei/mizukokudo_mizsei_tk2_000028.html (accessed 20 February 2022).

MLIT: website on the Current Status of Super Levees. *Study Meeting on Effective Construction of High Standard Levees. MLIT River Division*. The 1st Meeting on 18 May 2017. www.mlit.go.jp/river/shinngikai_blog/koukikaku_kentoukai/index.html Reference 2–1. www.mlit.go.jp/river/shinngikai_blog/koukikaku_kentoukai/dai1kai/pdf/2-1_genjyo.pdf (accessed 10 December 2020).

MLIT: website on High Standard Levees. www.mlit.go.jp/river/kasen/koukikaku/ What is High Standard Levees? www.mlit.go.jp/river/kasen/koukikaku/pdf/about.pdf (accessed 7 October 2021).

MLIT: website on Potential Inundation Areas and Hazard Maps. www.mlit.go.jp/river/bousai/main/saigai/tisiki/syozaiti/ (accessed 9 June 2021).

MLIT: website on the Tsurumi River. www.mlit.go.jp/river/toukei_chousa/kasen/jiten/nihon_kawa/0312_tsurumi/0312_tsurumi_00.html (accessed 7 December 2020).

MLIT: website on Typhoon Jane. www.kkr.mlit.go.jp/yodogawa/know/history/flood_record/flood_1950_shosai.html (accessed 6 June 2021).

MLIT KTR, Tokyo, Kanagawa and Yokohama (2007) *The Tsurumi River Management Plan*. MLIT Kanto Regional Development Bureau, Metropolitan Tokyo, Kanagawa Prefecture, Yokohama City, March 2007. www.pref.kanagawa.jp/documents/10592/tsurumi_seibikeikaku.pdf (accessed 31 May 2019).

MLIT Special Committee Recommendation (2015) *Disaster Prevention and Mitigation Responding to the New Stage*. MLIT Special Committee on Disaster Prevention and Mitigation Responding to the New Stage. January 2015. www.mlit.go.jp/common/001066501.pdf (accessed 17 December 2020).

Musiake, K., K. Ishizaki, F. Yoshino and T. Yamaguchi (1987) *Conservation and Recovery of Water Environment*. Sankaido Publishing Co., Tokyo.

NACSJ: website on Future of a Town Co-exist with Nature Sealed by Seawall (Minamisanriku-cho). www.nacsj.or.jp/2021/03/24743/ (accessed 10 November 2021).

Nagaoka City (2017) *Annual Report on Environment, 2019*. Section 2. www.city.nagaoka.niigata.jp/shisei/cate01/kankyou/file/h29-04.pdf (accessed 8 June 2021).

Nagasawa, Yasuyuki (1980) Underground Infiltration Method, Well Accepted! In: *Ienami*, No. 298, October 1980, Housing and Urban Development Corporation.

Nakagawa, Tomomi (2013) KOIZUMI Yakumo and HATTORI Ichizo. Lecture on the Yakumo Association General Assembly of FY 2013. *Herun (meaning Hearn)*, 51, 4–10. http://inamuranohi.jp/collection/%e3%80%8c%e5%b0%8f%e6%b3%89%e5%85%ab%e9%9b%b2%e3%81%a8%e6%9c%8d%e9%83%a8%e4%b8%80%e4%b8%89%e3%80%8d/#more-1373 (accessed 29 December 2021).

Namegaya, Y., K. Satake and S. Yamaki (2010) Numerical Simulation of the AD 869 Jogan Tsunami in Ishinomaki and Sendai Plains and Ukedo River-mouth Lowland. *Active Fault-paleo Earthquake Research Report*, 10, 1–21. www.gsj.jp/data/actfault-eq/h21seika/pdf/namegaya.pdf (accessed 13 November 2021).

NAOJ (2019) *Chronological Scientific Tables, 2020. National Astronomical Observatory of Japan*. Maruzen Co. Ltd., Tokyo.

National Astronomical Observatory (2021) *Chronological Scientific Tables, 2021*. Maruzen Publishing Co., Tokyo.

NIED: website on National Research Institute for Earth Science and Disaster Resilience. http://ecom-plat.jp/suigai-chikei/index.php?gid=10018 (accessed 31 May 2019).

Niigata Prefecture (2018, March) *Investigation Report on Environmental Impact on New Development of Water-solvable Natural Gas*. Division of Life and Environment, Niigata Prefecture. www.pref.niigata.lg.jp/uploaded/attachment/53200.pdf (accessed 18 February 2021).

NPA (2014, September 11) *Press Release on 11 September 2014*. National Police Agency Emergency Disaster Security Head Office.

NPA (2021, March 10) *Press Release on 10 March 2021*. National Police Agency Emergency Disaster Security Head Office. www.npa.go.jp/news/other/earthquake2011/pdf/higaijokyo.pdf (accessed 5 November 2021).

Oda, Yasunori (2008) *History of Public Nuisance and Environmental Problems*. Sekai Shiso-sha, Kyoto.

Okuma, Takashi (1981) *Transition of the Tone River Flood Control and Flood Disasters*. University of Tokyo Press, Tokyo.

Omachi, Toshikatsu (1998, September) *The River Law with Commentary by Article*. IDI Water Series 4. Infrastructure Development Institute-Japan, Tokyo.

Osaka Prefecture (2015) *The Neya River Block of the Yodo River System River Management Plan*. March 2015. www.pref.osaka.lg.jp/attach/4127/00011399/kasenseibikeikaku(neyagawa).pdf (accessed 8 August 2020).

Osaka Prefecture: website on Comprehensive Flood Control Plan. www.pref.osaka.lg.jp/ne/sougoutisui/gaiyo.html (accessed 8 August 2020).

Osaka Prefecture: website on Underground Rivers. www.pref.osaka.lg.jp/ne/sougoutisui/tikakasen.html (accessed 11 October 2021).

Osaka Sayama City Education Committee (2014) *Sayama Pond Comprehensive Academic Investigation Report*. www.sayamaikehaku.osakasayama.osaka.jp/oscm/report/syrp-043.pdf (accessed 23 January 2022).

Oya, Masahiko (1956) *Water-disaster Morphology Classification Map of Nobi Plain, Kiso River Basin*. http://ecom-plat.jp/suigai-chikei/index.php?gid=10018 (accessed 31 May 2019).

Peng, Yunyue and Kate Reilly (2021) Nature to Reshape Cities and Live with Water: An Overview of the Chinese Sponge City Programme and Its Implementation in Wuhan. *Report to EU Horizon 2020 Project GrowGreen* (Grant agreement No

730283).   https://growgreenproject.eu/wp-content/uploads/2021/01/Sponge-City-Programme-in-Wuhan-China.pdf (accessed 23 January 2022).

Reconstruction Agency (2016) *Housing Lots Provision Status for GEJE Refugees at the End of March 2016.* Issued 20 May 2016. www.reconstruction.go.jp/topics/main-cat1/sub-cat1-15/20160520_jutakukyokyuu.pdf (accessed 10 November 2021).

Reconstruction Agency (2021a) *Housing Lots Provision Status for GEJE Refugees at the End of December 2020.* Issued 12 January 2021. www.reconstruction.go.jp/topics/main-cat1/sub-cat1-15/material/20210112_jutakukyokyu.pdf (accessed 10 November 2021).

Reconstruction Agency (2021b, June 30) *GEJE Related Deaths.* www.reconstruction.go.jp/topics/main-cat2/sub-cat2-6/20210630_kanrenshi.pdf (accessed 5 November 2021).

Reconstruction Agency (2021c, October 29) *The Number of Evacuees in the Nation.* www.reconstruction.go.jp/topics/main-cat2/sub-cat2-1/20211029_kouhou1.pdf (assessed 6 November 2021).

Reconstruction Design Council (2011) *Towards Reconstruction: Hope beyond the Disaster 2011.* A Report Submitted to the Prime Minister by the Reconstruction Design Council in Response to the Great East Japan Earthquake on 25 June 2011. www.cas.go.jp/jp/fukkou/english/pdf/report20110625.pdf (accessed 6 November 2021).

River Council Interim Report (1977, June 10) On Comprehensive Flood Control Measures and Their Promotion Policy. In: Japan River Association (2004) *Collection of River Projects-Related Regulations.* Editorial supervision by Water and Disaster Management Bureau, MLIT. Japan River Association, Tokyo.

River Council Interim Report (2000a) *Effective Flood Control Including Basin Resistance.* 19 December 2000. www.mlit.go.jp/river/shinngikai_blog/past_shinngikai/shinngikai/shingi/ryuiki.pdf (accessed 14 August 2020).

River Council Interim Report (2000b) *Effective Flood Control Including Basin Resistance.* An Image of a Basin. 19 December 2000. www.mlit.go.jp/river/shinngikai_blog/past_shinngikai/shinngikai/shingi/T_03.html (accessed 12 December 2020).

River Council Report (1987, March 25) On Counter Measures Against Exceeding Floods and Their Promotion Policy. In: Japan River Association (2004) *Collection of River Projects-Related Regulations.* Editorial supervision by Water and Disaster Management Bureau, MLIT. Japan River Association, Tokyo.

Saito, Naohisa (1985) Recent Issues of Urban Rivers. *Textbook of the 21st Summer Seminar*, A-05, Hydraulics Committee, JSCE. http://library.jsce.or.jp/jsce/open/00027/1985/21-A05.pdf (accessed 15 February 2021).

Sankei News (2019) *Underground Tunnel to Drain Rainwater to the Osaka Bay.* Online news: The Sankei WEST on Typhoon 19, 2019.11.12 22:43. www.sankei.com/west/news/191112/wst1911120026-n1.html (accessed 8 August 2020).

Science Lab of the Daiichi High School (2000, September 13) The Nirasaki Volcanic Debris Flow Crossed the Basin. *Yamakai no Shiki*, 104. Daiichi High School, Kofu City. www.ypec.ed.jp/yamakai/yamakai%20104.html (accessed 3 December 2020).

SCJ (2019) *Tsunami Countermeasures for Nuclear Power Stations in Japan.* Science Council of Japan, Report. 21 May 2019. www.scj.go.jp/ja/info/kohyo/pdf/kohyo-24-h190521.pdf (accessed 5 November 2021).

Senga, Y., M. Sasaki, Y. Kawajiri and T. Endo (1981) Water Use Order. 3. Multipurpose Dams and Negotiation on Rivers. *JSIDRE*, 49(3), 235–243. www.jstage. jst.go.jp/article/jjsidre1965/49/3/49_3_235/_pdf (accessed 21 February 2022).

Shinano River Office: website on Okozu Bunsui. Hokuriku Regional Development Bureau, MLIT. www.hrr.mlit.go.jp/shinano/shinanogawa_info/naruhodo/suiro. html (accessed 4 December 2020).

Showa 24 Law No. 193 (1949) *Flood Fighting Law*. www.shugiin.go.jp/internet/itdb_ housei.nsf/html/houritsu/00519490604193.htm (accessed 19 February 2022).

Showa 31 Law No. 146 (1956) *Industrial Water Law*. https://elaws.e-gov.go.jp/ document?lawid=331AC0000000146 (accessed 19 February 2022).

Showa 32 Law No. 35 (1957) *Specific Multipurpose Dam Law of 1957*. https:// elaws.e-gov.go.jp/document?lawid=332AC0000000035_20200401_42 9AC0000000045 (accessed 23 January 2022).

Showa 33 Law No. 30 (1958) *Landslides Prevention Law*. www.shugiin.go.jp/internet/itdb_housei.nsf/html/houritsu/02819580331030.htm (accessed 19 February 2022).

Showa 37 Law No. 100 (1962) *Groundwater Pumping Regulation for Building Use Law*. https://elaws.e-gov.go.jp/document?lawid=337AC0000000100_20150801_ 000000000000000 (accessed 19 February 2022).

Showa 39 Law No. 167 (1967) *River Law*. www.shugiin.go.jp/internet/itdb_housei. nsf/html/houritsu/04619640710167.htm (accessed 23 February 2022).

Showa 44 Law No. 57 (1969) *Steep Slope Collapse Prevention Law*. www.shugiin. go.jp/internet/itdb_housei.nsf/html/houritsu/06119690701057.htm (accessed 19 February 2022).

Showa 45 Law No. 138 (1970) *Water pollution Control Law*. https://elaws.e-gov. go.jp/document?lawid=345AC0000000138 (accessed 19 February 2022).

Takagi, Fusetsu (1997) Efforts of Pioneers and Hanshin-Awaji Great Earthquake. *Preface, Natural Disaster Science*, 40(15–4), 251. www.jsnds.org/ssk/ ssk_15_4_251.html (accessed 10 February 2021).

Takahasi, Yutaka (1971) *Changes in Nation's Land and Water Disasters. Iwanami Shinsho, No. 793*. Iwanami Book Co., Tokyo.

Takahasi, Yutaka (1990) *Modern Japanese History of Civil Engineering*. Shokoku-sha, Tokyo.

Takahasi, Yutaka and Toshitsugu Sakou (1963) *History of Japanese Civil Engineering*. Chijin Shokan, Tokyo, 1960.

Takasaki, Tetsuro (1996) *River Flows in Desert: 500 Days to Save the Great Tokyo Drought*. Diamond Publ. Co., Tokyo.

Takeuchi, K. and S. Tanaka (2016) Recovery from Catastrophe and Building Back Better. *JDR*, 11(6), 1190–1201. http://doi.org/10.20965/jdr.2016.p1190.

Takeuchi, K. and S. Tanaka (2021) Anticipated Maximum Scale Precipitation for Calculating the Worst-Case Floods. *Water Policy*, 23(S1), 128–143. http://doi. org/10.2166/wp.2021.241.

Tamagawa Sampo: website on the Monument of the Tama River Breach. http:// tamagawa.circlemy.com/history-02.html (accessed 8 December 2020).

Taniguchi, Mitsuomi (2006, April 28) *Review of the Daitoh Flood Suit*. Lecture at the Meeting of Love Rivers in Osaka. www.japanriver.or.jp/circle/oosaka_ pdf/2006_taniguchi.pdf (accessed 8 December 2020).

Thompson, Christopher (2020) "Inamura no hi" ("the Rice Bale Fire"), Its Evolving Story and Global Relevance: The Politics of Tsunami Preparedness in Japan. Article in *Disaster Prevention and Management*, October 2020. http://doi.org/10.1108/DPM-07-2019-0211.

UNESCO IHP RSC: website on RSC for Southeast Asia and the Pacific. http://hywr.kuciv.kyoto-u.ac.jp/ihp/rsc/index.html (accessed 4 March 2022).

Wakabayashi, Takako (2015) *Musashi Canal to Be Reborn*. www.water.go.jp/honsya/honsya/pamphlet/kouhoushi/2015/pdf/201508_06.pdf (accessed 4 December 2020).

Wong, Tony H.F. (2006) An Overview of Water Sensitive Urban Design Practices in Australia. *Water Practice & Technology*, 1(1). http://doi.org/10.2166/wpt.2006.018.

Yamaguchi Prefecture: website on Typhoon Ruth. www.pref.yamaguchi.lg.jp/cmsdata/a/1/d/a1d354beb3a334620491e21ce856ac01.pdf (accessed 7 June 2021).

Yamamoto, Saburo and Shigeki Matsuura (1996, October) Establishment of Old River Law and River Administration (2). *Suiri Kagaku*, 231, 51-78. https://agriknowledge.affrc.go.jp/RN/2010550264.pdf (accessed 13 January 2022).

Yamanashi Kanko: website on Sansha Shrine. www.yamanashi-kankou.jp/kankou/spot/p1_4327.html (accessed 16 January 2022).

Yamanashi Prefecture (2005) *The Kamanashi River Region of the Fuji River System Improvement Plan*, October 2005. www.pref.yamanashi.jp/chisui/documents/80834707320.pdf (accessed 29 May 2021).

Yamashita, Fumio (2008) *Tsunami and Disaster Prevention-Sanriku Tsunami Stories*. Kokon Shoin Publishing Co., Tokyo.

Yanagida, Kunio (1975) *Empty Weather Map*. Shincho-sha, Tokyo.

Yasui, Masahiko (2011) A Study on a Civil Engineer Takashi Mizutani and His Works in the Multiple-purpose Works on River Basin in the Early Stage. *Proc. of JSCE*, D2 67(1), 21–37. www.jstage.jst.go.jp/article/jscejhsce/67/1/67_1_21/_pdf (accessed 21 February 2022).

Yodo River Office: website on the History of the Yodo River. Kinki Regional Development Bureau, MLIT. www.kkr.mlit.go.jp/yodogawa/know/history/now_and_then/tanjyou.html (accessed 4 December 2020).

Yodo River Office: website on the Yodo River Reform. Kinki Regional Development Bureau, MLIT. www.kkr.mlit.go.jp/yodogawa/know/history/now_and_then/kouji.html (accessed 4 December 2020).

Chapter 6

Future issues of IFRM

Natural as well as social conditions where disasters occur continuously change over time. Current global changes in hazards, exposure as well as the vulnerability of society are so profound and rapid that the impacts are difficult to anticipate. But it is certain that climatic change is making hydro-meteorological extremes amplified, population increase and urban concentration are making people's exposure to flood and sedimentary disaster risk areas increased, and societal complexity, dichotomy of rich and poor and loose tie among community members are all making societal vulnerability to disasters higher. The efforts of nations and the UN are serious and appreciated but the tendency of increasing risk seems not stopped nor weakened but accelerated.

How does the Earth change? Where does society go? How should people do? They are questions of global nature, population, civilization, and individual people. There is yet no answer but there are some indications. This chapter briefly follows the reality and highlights the social capacity and transdisciplinary approach.

6.1 ANTICIPATED CHANGES AND ADAPTATION

6.1.1 Climatic changes

Changes in temperature and heavy rains

The cause of current climatic change is considered the increase of greenhouse gasses and a resultant global warming. This changes the sea surface temperature and global climatic system including El Nino-Southern Oscillation (ENSO), Indian Ocean Dipole (IOD), North Atlantic Oscillation (NAO), westerlies, monsoons, etc. They change seasonality, size and locations of high pressures, low pressures and cyclones. Heavy rains, changes in snowfalls, super cyclones, increased variation of seasonal and

DOI: 10.1201/9781003275541-7

annual precipitation and prolonged droughts may be the main forms of climate change.

The salient features of Japanese climate change are high temperature, torrential rains, linear rain bands, large-scale typhoons, etc. Figure 6.1(a)–(d) show chronological changes of temperature and precipitation measured by the Japan Meteorological Agency (JMA) Observatories since June 1875 and the Automated Meteorological Data Acquisition System (AMeDAS) started on 1 November 1974 (JMA: website on JMA history).

Figure 6.1(a) shows the number of days over 35°C air temperature per year per station over 13 stations since 1910 and Figure 6.1(b) the number of days over 200 mm/d per year per station over 51 stations since 1900, both consistently observed stations. Figure 6.1(c) is the number of days over 400 mm/d per year per 1300 stations, and Figure 6.1(d) the number of days over 50 mm/h per year per 1300 stations, both based on 1976–2019 AMeDAS records (JMA: website on Changes in Extremes). Note that although the number of AMeDAS stations is now about 1300, they were less when it started in 1974 until March 1979 when the number of stations reached 1316 so that the total number of days observed in different years were adjusted to 1300 stations.

Both temperature and heavy rains show increasing trends. In JMA's 13 stations, the temperature increase in Figure 6.1(a) is remarkable, especially since the early 1990s. In Figure 6.1(b), the increase of the number of days over 200 mm/d is also significant over 100 years but seems rather stable in the period of 1940–1980 and again increasing in the recent 40 years. From the recent 44 years (1976–2019) of AMeDAS data, both the number of days of heavy rains over 50 mm/h in Figure 6.1(c) and over 400 mm/d in Figure 6.1(d) show a clear increase. While over 400 mm/d cases are remarkable, however, over 50 mm/h cases show two modes, low and high, and the high mode seems not clear in increasing trend in the past decade and a half.

The two mode phenomena of Figure 6.1(c) may be due to the number of heavy rain events, many or few, in a year. For example, in 2014, record-breaking 10 typhoons landed and 19 approached Japan and had many rain events that increased the number of days with over 50 mm/h rains (JMA: website on Typhoon Statistics). In contrast, in 2019, the number of typhoons that landed in Japan was 5 and approached was 15 but Typhoon 19 was a super typhoon that brought the record-breaking rainfall in large part of East Japan in only 4 days as shown in Figure 6.4 which made the number of days with over 50 mm/h not so many but the days over 400 mm/d became the top in JMA's recorded history.

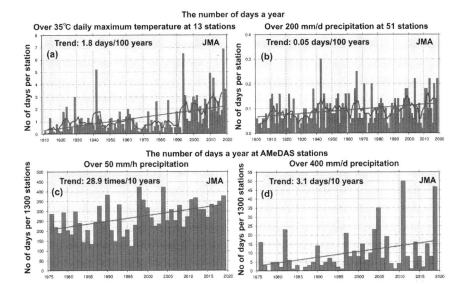

Figure 6.1 Climate change in temperature and frequency of heavy rains. (Courtesy of JMA, JMA: website on Changes in Extremes)

(a) Number of days over 35℃ days per year per station over 13 stations.

(b) Number of days over 200 mm/d per year per station over 51 stations.

(c) Number of days over 50 mm/h per year per 1300 stations.

(d) Number of days over 400 mm/d per year per 1300 stations

Thus, it may be concluded that the climate change impact on heavy rains is clear in intensification over a long time period but with considerable fluctuations in short periods.

Table 6.1 lists the frequency of occurrence of the top 20 records of air temperature, 10-minutes rainfall, 1-hour rainfall and 1-day rainfall in the periods: before 1990, 1991–2000, 2001–2010 and 2011–2020 observed at AMeDAS stations since 1976 plus some other JMA observation stations where long term continuous and consistent records were available (JMA: website on Historical Ranking of Extreme Records).

It is remarkable to see that so many record breakings occurred in the latest decade 2011–2020, all top six in air temperature, top two in 10 minutes rainfall and 1-day rainfall. But with respect to 1-hour rainfall, only the fourth was in. Such a phenomenon is quite interesting but difficult to explain. But once again, it gives a caution that the overall phenomena cannot be seen from a partial data not only in time series but also on the phenomenal items. It is not only misleading but also possible to manipulate

Table 6.1 Frequency of Occurrence of the Top 20 Extremes in AMeDAS Records Since 1976 and Some Other JMA Stations*. (JMA: website on Historical Ranking of Extreme Records)

	Before 1990	1991–2000	2001–2010	2011–2020 (top 10 ranking**)
Air temperature	1	3	3	13 (1–6,8,9)
10-min. rainfall	11	2	0	7 (1,2,6,9)
1-hour rainfall	8	3	4	5 (4)
1-day rainfall	6	2	5	7 (1,2,7)

* Continuously and consistently recorded stations with different data length where the oldest data ranked in was 1933 for temperature, 1937 for 10 minutes, 1944 for 1-hour and 1976 for 1-day precipitation.

** The ranking of top 10 records is shown for the latest decade 2011–2020.

the reality to a false understanding by showing a partial aspect of the total data.

Linear Rainbands

Linear rainbands (senjo-kousuitai) are the rainfall phenomena that recently occurred quite often resulting in serious disasters. According to Kato (2020), there are two categories of linear rainbands:

> the broken line type in which convective cells simultaneously form on a quasi-stationary local front by the inflow of warm and humid air, and the back building type in which new convective cells successively forming on the upstream side of low-level winds linearly organize with pre-existing cells.

In simple words, the broken line type forms several heavy rain areas simultaneously online and each stays there for a while. The back building type forms a heavy rain area successively one after another at an upstream point and each flows down online towards downstream. To follow are two linear rain bands that brought serious damages.

The August 2014 Heavy Rain

In August 2014, typhoons 12 and 11, and the frontal activities caused record-breaking rainfalls in many places from Kyushu to Hokkaido which was named by JMA "August 2014 heavy rain". In Hiroshima, the back building type linear rainbands brought hourly rainfall of 121 mm and 3-hourly rainfall of over 200 mm at Asa-kita station on 20 August that caused several lines of severe landslides and killed 74 people in Hiroshima prefecture. (JMA, 2014; Hiroshima Prefecture: website on 20 August 2014 heavy rain disaster).

The 2018 July Heavy Rain (West Japan heavy rain disaster)

The Baiu front stagnated in northern Japan since 28 June 2018 came down to western Japan by 5 July and stayed around. Typhoon 7 originated on June 29 in the southern ocean of Japan passed from the East China Sea to the Sea of Japan where it became an extratropical cyclone on 4 July. The front and the typhoon kept supplying warm humid air and brought heavy rains all over Japan, especially in western Japan. The 10-day rainfall from 28 June to 8 July exceeded 1800 mm in Shikoku, 1200 mm in Tokai and 900 mm in northern Kyushu. In many stations, record-breaking heavy rains were observed such as 1-, 3-, 6-, 12-, 24-, 48- and 72-hours historical rainfall records were renewed at 14, 16, 31, 48, 76, 124 and 122 stations out of about 1300 AMeDAS stations in Japan (JMA, 2018). It was amazing that nearly 6% of all stations in Japan renewed their 1-day rainfall records and nearly 10% of stations renewed their 2–3 days records in just one rainfall event. Such long-lasting rainfall extremes caused serious floods and land-slides which resulted the dead and missing 237 and 8 (FDMA, 2019) in entire Japan. A hydro-meteorological disaster with over 200 casualties was since the 1982 Nagasaki disaster in which 299 died (CO: website on Naga-saki Water Disaster).

During this July 2018 heavy rains, back building type linear rainbands were formed at 15 places in western Japan and Heavy Rain Special Warn-ings (see Table 5.3) were issued in 11 prefectures. This included Hiro-shima again and its western adjacent prefecture Okayama as shown in Figure 6.2.

In Aki and Asa-kita, Hiroshima City, over 480 mm/4 days, and over 60–70 mm/h rain fell and severe debris flows occurred resulting dead and missing 23 and 2 in Hiroshima City (Hiroshima City: website on the outline of the July 2018 heavy rain disaster). In Okayama prefecture, over 300–450 mm/3 days of rain fell in many stations and floods and debris flows occurred which killed 61 people in Okayama prefecture. Especially in Mabi town at the confluence of the Oda River flowing into the main Takahashi River, a major flood occurred by breaches of levees of the Oda River and its branches at six sites due to high water of the Takahashi River flew as backwater into the Oda River. In Mabi town alone, 51 people died, of which more than 85% were over 65 and died at home

Figure 6.2 Linear rainbands over western Japan during the July 2018 heavy rain reported by NHK TV. (Courtesy of JMA, NHK on 7 July2018 at 0:41 am)

(Okayama prefecture: website on the Post Audit Report of the July 2018 Heavy Rain Disaster).

East Japan Typhoon (Typhoon 19, 2019)

Typhoon 19 (Hagibis), 2019 was an extra-large typhoon landed on the Izu Peninsula on 12 October 2019 with 955 hPa and, having passed through Kanto and Tohoku regions, left to the Pacific Ocean on 13 October. Figure 6.3 compares its satellite image with that of a normal size Typhoon 15 that landed on Chiba Prefecture just a month ago and caused a serious wind disaster in southern Kanto. With such incomparably large scale clouds, their extreme impact to Japan was obvious. As shown in Figure 6.4, typhoon 19 was truly a super typhoon brought much rainfall over a large part of Japan from Kinki to Tohoku regions especially Shizuoka, Niigata, Nagano, Yamanashi, Kanto and Tohoku regions. In Hakone, 1001.5 mm and in 17 stations in East Japan, over 500 mm were observed from 10–13 October. "Heavy Rain Special Warnings" were issued in 13 prefectures and caused floods in many rivers in East Japan (JMA, 2019). The dike breaches happened at 142 sites in 71 rivers resulting in large damages with dead and missing 107 (including disaster-related deaths) people of which two-thirds were concentrated in Fukushima, Miyagi and Chiba prefectures (JSCE, 2020; MLIT, 2019a). It was named the East Japan Typhoon by JMA. Still worse only about 10 days later, on 24–26 October low pressure passed from

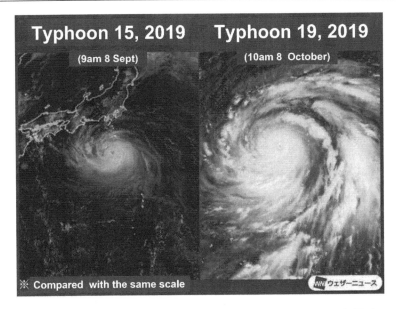

Figure 6.3 Large scale Typhoon 19 of 2019 in comparison with ordinary scale Typhoon 15 of 2019 in the same scale. (Courtesy of Weathernews and NICT, Weathernews: website on Typhoon 19 is greater than Typhoon 15)

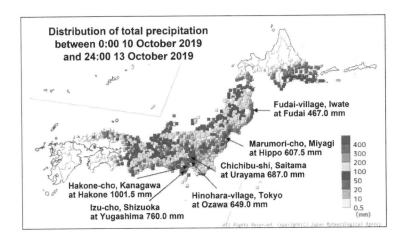

Figure 6.4 The total rainfall from October 10 at 0:00 to October 13 at 24:00 by Typhoon 19, 2019. A large part of eastern Japan received much rain because of super-scale typhoon and serious floods everywhere from Chugoku to Hokkaido. (Courtesy of JMA, JMA, 2019)

western Japan to northern Japan stimulated by Typhoon 21 and brought again much rain resulting in another 13 people's deaths (CO, 2020).

The Chikuma River, the Nagano part of the Japanese longest river Shinano, had several major dike breaches by which many houses were inundated. An example is shown in the front cover photo of this book which was the 70 m dike breach in the left bank of the Chikuma River at Hoyasu of Naganuma District, Nagano City on 13 October 2019. By this breach, even ten Shinkansen trains (120 cargoes) were inundated in the Hokuriku Shinkansen train base at Akanuma of Naganuma District. (Nagano City, 2021). This became once again a strong alert for the necessity to follow the indication of hazard maps.

The Abukuma River is a 5400 km² basin with a river length of 239 km stretching from Fukushima to Miyagi prefectures, that is, from south to north (MLIT: website on the Abukuma River). The basin shape is like a feather with a length of 125 km and average width of 40–45 km. Because it is flowing from south to north, the typhoon-induced rains followed the flood flow from upstream to downstream which made the mainstream river flood increasingly high and caused backwater into many branch rivers joining to the mainstream. This is why among all 52 dike breaches occurred, 45, more than 85%, were in the branch rivers (CO, 2020).

Typhoon 19 was quite similar to 1958 Typhoon Kanogawa in scale, power and the route. Typhoon Kanogawa had a much smaller rainfall area mainly only in the Kanto region but caused a huge damage in the Kano River basin where dike breaches occurred at 14 sites, inundated 6775 houses and the death tolls reached 853. By Typhoon 19, however, no dike breach occurred in the Kano River basin and none died although about 1300 houses were inundated by inland floods. It was helped by the Kano River Diversion Canal completed in 1965 which functioned perfectly (MLIT Numazu River and Road Office, 2019).

6.1.2 Adaptation to climatic change

Thus the climatic change is steadily progressing with global warming and intensification of hydro-meteorological extremes. Its impacts are profound. Increase of air temperature results in sea level rise that increases coastal retreat and flood risk, move of climatic zones, in general, from south to north that changes geographical distribution of biota, brings early snow melt, decrease of snow storage, etc. The change in climatic zone would be a long-term process but the intensification of extremes such as floods and droughts is very rapid and already taking place in various ways as seen in Japanese cases in the previous section. The eventual destination in the future stationary state is unknown with unknown societal changes. Even with conditional changes under a certain societal scenario, it is still uncertain especially in local-scale phenomena despite tremendous efforts such as IPCC-related research. Therefore, climate change adaptation is necessarily a process of decision-making under uncertainty. Under such conditions, the response strategy should be adaptive, robust, flexible and, above all, precautious to unknown changes and disturbances to come.

Figure 6.5 shows a necessary climate change adaptation process for flood risk management. In any nations with severe natural hazards such as the Asian monsoon region or the Pacific-rim and Hindu Kush Himalayan (HKH) orogenic zones, the current infrastructure level is not at all satisfactory and in the process of improving towards current tentative targets, not yet for the final target. In the figure, the tentative target on flood safety is against non-exceedance probability 1/150 or 150-year floods. But the current safety level achieved is only up to a probability of 1/70 or so. The intensification of hydro-meteorological phenomena decreases even such the currently achieved low safety level 1/70 to only 1/20 or so and the tentative goal 1/150 to only 1/40 or so. In order to achieve the original target of 1/150 after climate change, much further effort is necessary. Therefore, in many nations, it is impossible to achieve the new target with structural infrastructure and major efforts are necessary to augment the non-structural means such as land use regulation, early warning and evacuation, various preparedness supposing that extreme disasters happen. Such a difficult situation, that is, while the current target is yet little achieved, the target itself goes far, is especially prominent in the nations under intensive natural hazards. Hazardous

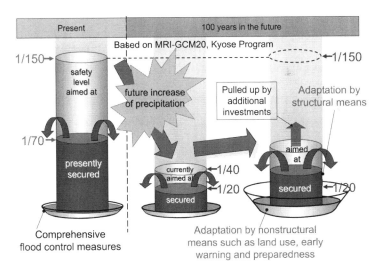

Figure 6.5 Climatic change impact on flood risk and adaptation strategy. In flood prone countries, the current safety target aimed at under present climate has been far from achievement so that climate change is an extra burden and human adjustment by nonstructural means becomes more important. (Courtesy of MLIT, MLIT, 2008)

nations become even more hazardous and the efforts required must become tremendous.

6.1.3 Demographic issues

Other than climate, there are many other sources of anticipated difficulty and uncertainty in the future. Among them the political unrest that leads to conflicts, civil wars and wars would be the most serious ones which determine nation's fate. Economic and other socio-cultural disturbances would also be high uncertainty that influence living conditions and safety. Among them, a relatively certain future change would be the demographic one. In general, it would be population increase and urbanization. Both tend to increase residents in hazardous areas and activities vulnerable to hazards. They would be the most certain threat in many nations in the world. But in nations such as Japan, the Republic of Korea, China as well as many European nations, the issues are aging and depopulation, especially in rural areas that create new risks. The potential remedy in any case would be science and technology, and governance. This section touches upon issues due to demographic changes.

Urbanization

Although the world population is expected to keep growing this century, the situation is different nation by nation. There are quite a few countries where depopulation has already started, and even in a region-wise, as seen in Table 6.2, the European population slightly declines by 2030 and continues further in the century. Similarly, Asian and Latin America and Caribbean populations are expected to decline between 2050 and 2100.

Table 6.2 Population Projection of the World and Regions, 2017, 2030, 2050 and 2100. The Medium-Variant Projection of UN DESA Population Division (2017).

Region	Population (millions)			
	2017	2030	2050	2100
World	7 550	8 551	9 772	11 184
Africa	1 256	1 704	2 528	4 468
Asia	4 504	4 947	5 257	4 780
Europe	742	739	716	653
Latin America and the Caribbean	646	718	780	712
Northern America	361	395	435	499
Oceania	41	48	57	72

Population projection involves difficult factors and according to different fertility, mortality and migration assumptions, there are several projections. For urban population projections, the median of various population projections (UN DESA Population Division, 2017) is used, and multiplied by the trends of the proportion of the population living in urban areas (UN DESA Population Division, 2019).

As seen in Figure 6.6, urban population ratio increases by economic growth. Urbanization is a positive factor for economic growth by concentration of business entrepreneurs, quality labors, infrastructure, etc. and economies of scale.

Depopulation and aging society

As seen in Table 6.3, in all nations, including where depopulation is taking place, the proportion of urban population is increasing. In Japan, where not only total population but also urban population is projected to decline by 13.5 million between 2018 and 2050, the proportion of urban population still increases from 92 to 95%. In the Republic of Korea and Germany where nations' total populations decline, their projected urban populations slightly increase by 0.1% (375,000) and 0.3% (2955,000) between 2030 and 2050 but in the proportion wise, they increase as much as 4% and 5%, respectively. It simply reflects that urban concentration continues while depopulation is taking place in rural areas as well as in the whole nation.

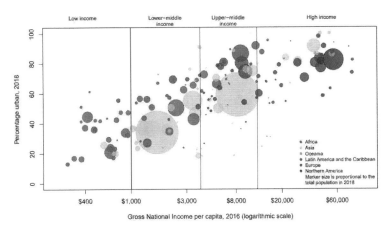

Figure 6.6 Urbanization rate versus gross national income (GNI) per capita. (UN DESA Population Division, 2019)

Table 6.3 Urban Population Projections (UN DESA Population Division, 2019).

	Urban population (thousands)				Urban population (%)				Av annual change %
	1990	*2018*	*2030*	*2050*	*1990*	*2018*	*2030*	*2050*	*2015–2020*
Japan	96,299	116,522	112,710	103,039	77	92	93	95	−0.1
Indonesia	55,491	147,603	185,755	234,105	31	55	63	73	2.3
Viet Nam	13,817	34,659	47,248	65,711	20	36	44	57	3.0
Philippines	29,106	49,962	63,844	93,465	47	47	51	62	2.0
Malaysia	8,982	24,364	30,109	36,440	50	76	82	87	2.1
China	310,022	837,022	1,017,847	1,091,948	26	59	71	80	2.4
R. Korea	31,696	41,678	43,241	43,616	74	81	82	86	0.3
Germany	57,850	63,622	64,871	66,826	73	77	79	84	0.3
Italy	38,119	41,763	43,161	44,671	67	70	74	81	0.3
Ukraine	34,356	30,521	29,537	28,634	67	69	72	79	−0.3
USA	190,156	268,787	301,001	347,346	75	82	85	89	0.9

Urban growth and rural depopulation are both definite sources of increasing flood disaster risk and a full consideration of IFRM is necessary. Furthermore, aging society multiplies the problem. Decline of population, especially decline of the young people means the decline of people who can support flood management, especially those who can help vulnerable people. Preparedness would become limited by less participation of young people and infrastructure installment becomes more costly as economy of scale is lost. It is already a profound problem emerging in many depopulated rural areas. What could be a solution to such a problem?

6.1.4 Compact city

One solution to the problem of depopulation and aging society is the concept of compact city. It is not only for depopulation and aging but also for many other contemporary societal issues such as disaster risk, energy, environment, social capital, etc. Therefore it applies to population increasing nations, too. It is considered as *an ace in the hole* for economic efficiency, energy-saving and environmental protection against urban sprawl and sparse rural habitation.

Compact city is characterized by the following conditions (OECD, 2012):

(1)   City is developed with dense and proximate buildings and facilities.
(2)   People have high accessibility to local services and jobs.
(3)   Each city is connected by public transportation.

The first condition "dense and proximate buildings and facilities" is the key to the compact city. Because of this closeness, people can access to services and jobs easily without much travel or automobile use. This results in little energy use and efficient land use which contribute for economic efficiency and environmental protection.

The background problems of the urban sprawl and rural depopulation may be summarized as follows:

(1) Urban sprawl destroys natural environment and biodiversity. As urbanization continues and urban areas grow larger into the suburbs that is a pressure to the surrounding nature.
(2) Urban sprawl and rural depopulation make energy efficiency low and CO_2 emission high.
(3) Urban sprawl and rural depopulation make protection from natural hazards difficult and costly to develop protective infrastructure.
(4) Urban sprawl and rural depopulation make quality of life difficult as provision of various utilities and services such as roads, transportation, water, energy, health, daycares, etc. to the entire population become costly and inefficient.
(5) Residential houses and retail shops tend to move from the densely inhabited district (DID) to suburbs (doughnut phenomenon) making the density of urban areas lower and city life inconvenient.
(6) Especially as many elderly people live in the suburbs or depopulated rural areas, to keep good accessibility to health care and other services for them becomes difficult.
(7) Low density of residents makes the community communication and activities difficult and degrade social capital.
(8) Small isolated villages in rural area have all problems of inefficiency in provision of energy, utility, services, healthcares, education, safety and especially with aging, problems become serious.

In short urban sprawl and sparse rural habitation make accessibility to and provision of any utility, services and safety time and energy-consuming, economically inefficient, and environmental quality degrading.

The idea of compact city is to solve all those problems at once. As depicted by the schematic view of Figure 6.7, the concentrated formation of a city makes time and space distance small, energy efficiency high, economies of scale high and pressure to the nature small. Especially for disaster risk management, it makes it possible to build strong infrastructure to protect only small area and give the rest of area free to any nature including flood retardation. Question is only how to realize such compact cities. There are many obstacles such as consensus building, creation of new lifestyle, concrete procedure to shift existing cities to compact cities etc. It will be a slow process and needs strong and long-term advocacy, learning and

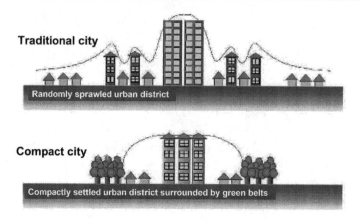

Figure 6.7 Basic concept of compact city (Study Team for In-town Residence, 2000). (Courtesy of MLIT)

above all, political leadership with strong commitment. Efforts of MLIT Tohoku Regional Bureau would be one of a few precious examples (Study Team for In-town Residence, 2000; MLIT, 2015; MLIT Urban Bureau, 2017; OECD, 2012).

Case of Japan

Japanese population peaked in 2008 at 128.084 million and gradually decreased (IPSS, 2012). It was 125.57 million in January 2021 (MIC, 2021). According to the National Institute of Population and Social Security Research's 2012 report (IPSS, 2012), in 2050 the Japanese population median projection is 97.1 million and in 2100, 49.6 million. Such a sharp decline! But the projection largely varies with the birth and death rate estimates by each national census. According to the 2017 report, the estimates in 2050 and 2100 are 101.9 and 59.7 million (IPSS, 2017), a considerable difference.

Although the projections are uncertain, urban sprawl, doughnut phenomenon (urban residents move from the urban center to the suburbs), rural and suburbs depopulation have been rapidly progressing and suburban disaster risk increasing, that cannot be left overlooked. To address such issues the Japanese Parliament passed a law "Urban Revitalization Special Law" in 2002 (Japanese Law No. 22, 2002) which was continuously amended to reflect the urgent needs for sparse population in the suburbs, doughnut phenomenon and disaster safety of the city. In the law, the methodologies of compact city development are indicated that CTVs are suggested to make "urban settlement adjustment plan" which identifies "residence promoting zone" and "urban function promoting zone" as shown in Figure 6.8. In order to concentrate residents and urban facilities in low disaster risk area, those districts are instructed not to include disaster red zone.

The MLIT has been promoting compact city to improve the decline of business and living conditions due to urban sprawl and depopulation in the suburbs not by a piecemeal prescription but under a more fundamental strategy. Their philosophy is to make urban economy and finances sustainable, elderly's life and children's nursery conditions improved, energy and the natural environment sound and disaster safety improved (MLIT, 2015). The basic approach is a slow but continuous effort to shift the current centralized state to a multi-node decentralized state. The basic methodology is to support the Urban Revitalization Special Law of 2002 and its amendments, that is, promoting establishment of "urban settlement adjustment plan" with designating residence and urban facilities concentration districts (Japanese Law No. 39, 2014). Furthermore, included is the discouragement

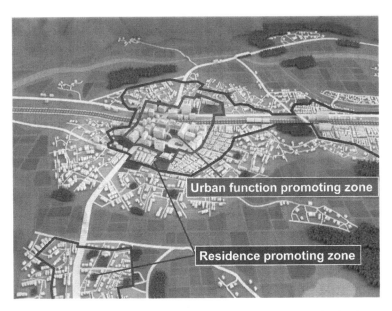

Figure 6.8 A schematic illustration of urban settlement adjustment plan by designating residence promoting district and urban function promoting district where disaster red zone should not be included (MLIT: website on Compact City Promoting Project). (Courtesy of MLIT)

of the new residence in "urbanization control area" introduced by the City Planning Law of 1968 (Japanese Law No. 100, 1968). For example, the exceptional permission of development given in urbanization control area by Item 11 of Article 34 of this law of 1968 is now reconsidered to stop urban sprawl. An MLIT recent review reported that "the settlement adjustment plan" was prepared by 250 CTVs while 468 CTVs started preparation out of the total 1374 CTVs in Japan as of the end of 2018 (MLIT, 2019b). The idea is slow but steadily progressing.

6.2 IMPORTANCE OF SOCIAL CAPITAL

Disaster preparedness is the key to reduce disaster losses, especially human lives. Preparedness can be realized by combined efforts of the public, community and individuals. Roles of community and individuals are particularly important at the beginning of a disaster as when a disaster occurs the immediate action of local people on spot controls the fate of disaster afterwards. The disaster losses are much suppressed by individuals' behavior properly instructed by local leaders. In order to make it happen, daily exercises of local people such as regularly scheduled drills are necessary. Such exercises may be organized and promoted by an administrative order but more effectively with disaster preparedness culture. This section discusses how local disaster preparedness culture may be created and kept working in local communities.

6.2.1 Social capital and its recent decline

Social capital is the value of social networks to effectively work together to achieve a common objective. It is a capacity of human network of society to absorb the hardship, share the well-being and safeguard the weak members within society by mutual help with or without institutional arrangements. Such network is typically formed among family members and neighbors in a community and serves as a societal bondage to mutually safeguard the collective well-being.

In disaster risk reduction, it is nearly a synonym to mutual help in a community to save themselves from the devastation of a disaster. But it applies not only for a disaster occasion but also any occasion when anybody in a community falls in trouble, temporally or permanently, such as crimes, accidents, family bankruptcy, etc., especially involving the handicapped, the aged, infants and other vulnerable people who need help when they encounter difficulties.

For preparedness and emergency responses including evacuation exercises, for example, a network of mutual help is the heart of disaster loss reduction. Even an earliest public help can come after some time, and while waiting, what is available are only self-help and mutual help. Even after it arrives, quite often it can be executed through the mechanism available for mutual help. A network of mutual help is therefore the heart of preparedness.

Then, how can it be effectively formed? It is, in fact, a question of how to form disaster preparedness culture.

Disaster preparedness culture

In Japan, routine exercises of emergency preparedness are popular and well-practiced in local communities. In many elementary schools their playgrounds serve as temporary retardation ponds and pupils learn flood mechanism and receive evacuation training. In many flood-prone areas, evacuation exercises and dike watching and repairment training are also regular exercises. In landslide and debris flow-prone areas, evacuation training is legally compulsory for all residents in designated communities in yellow zone.

Figure 6.9 illustrates example pictures of dike protection training held at various rivers in Japan before the flood season starts, normally in the

Figure 6.9 Upper and left bottom: Flood fighting training held in the left bank riverbed of the Fuji River at Kijima, Fuji City, Shizuoka Prefecture on 5 June 2019 participating 38 members organized by the Disaster Mitigation Council of the Fuji River Basin (Photo: Courtesy of Kofu River Road Office, MLIT). Right bottom: Dike protection training on *Tsukinowa* method. (Photo: Courtesy of MLIT)

"flood prevention month" designated in May except Hokkaido in June. The frontline players of local flood disaster prevention are, historically, local governments (CTVs) which form flood-fighting management associations and jointly work with community flood-fighting teams. Most flood-fighting teams are fire brigades and if a fire brigade cannot cover floods, a separate flood-fighting team is organized. There were 3252 flood-fighting management associations nationwide in April 2001 (CO: website on Flood Fighting Team), and 2171 fire brigades and 71 flood-fighting teams, altogether about 870,000 brigade or team members as of April 2016 (MLIT, 2018). They are volunteers with their own job. Therefore it is difficult to recruit and keep young volunteers in the team. They are expected to engage in such works as watching dikes; flood protection engineering such as protecting dikes by putting sandbags on the top, covering sheets or trees with many leaves (*Kinagashi*), stopping infiltration by making a pond outside of the dike (*Tsukinowa* method); helping evacuation and rescue operations; fixing drainage works, etc.

For such community preparedness exercises, human network of mutual help is vital for success. But an efficient network of mutual help cannot be formed at once by an administrative order or a top-down instruction but as a result of long-term history of daily life, that is, a disaster preparedness culture.

Awareness and consciousness of risk in daily life make communities to take a collective action for emergency response and develop mutual help during and after a disaster. They form disaster preparedness culture. But such culture is formed not only by disaster preparedness activities but also by exercises of any local activities such as festivals, sports events, cleaning, recycling or even funerals where people work together for community. Radio gymnastics at elementary school or a nearby community space have long been one of the popular exercises of community gathering. Seasonal festivals with miniature shrines have historically been the most popular occasion of community gathering. One of them, depicted in Figure 5.3 in Section 5.2.1, is the Omiyuki-san festival in Ryuo, Kai City, Yamanashi Prefecture has been continuing that started specifically to pray for safety from a flood disaster and later to jointly protect the Shingen Bank.

Unfortunately, however, such tradition or occasions of community gathering is now declining anywhere in the world, especially where urbanization is taking place. This is due to high labor mobility so that they:

do not live long in the same place,
do not live with multi-generations,
do not participate and maintain local festivals,
do not necessarily work nearby,
do not pay much attention to their neighbors, and
live rather independently with their own interest.

Under such conditions, community cannot maintain and accumulate what they learned in disaster and other emergency experiences. The experienced people move out replaced by new people who have no experience of disasters in their localities. The elderly's knowledge cannot be succeeded to their children as they do not live together. There are few occasions to meet and know neighbors each other as events of community gathering are getting fewer. Furthermore, a majority of people work far out of their residential community and establish their main human connection at job site rather than a residential site.

In Japan, emergency response is largely shared by fire-fighting volunteer groups of local communities who do not work only for fires but also for any disasters such as floods, landsides, earthquakes or any accidents such as appearance of wild animals. Their function has been very important but the number of volunteers are decreasing because the number of young people who stay and work nearby their residential community is decreasing. Rural depopulation, urban doughnut phenomenon and urban sprawl are accelerating the situation. This tendency is all over Japan and the volunteer groups are now being replaced by hired construction companies around the community.

Thus disaster culture has value that constitutes social capital but is declining due to urbanization, labor mobility and other changes in lifestyle. The basis of traditional human relation is disappearing one way or another. While such tendency may be reconsidered under the concept of sustainable society, it does not mean to go back to traditional society. A key question is whether decline of social capital can be replaced by information technology, cell phones and various social network services. Social communication media and network have changed the sense of distance wherever may be, and the big data show a great potential of grasping the flow of people and analyzing it for social guide. That is true. But nevertheless, as human being is not a robot, social capital and disaster preparedness culture can never be completely or even considerably replaced by information technology. The traditional societal culture would remain important for a community to remain safe and trustable each other to live life together and raise future generations. The following example of keeping social capital over a thousand of years would be a lesson worthwhile to see.

6.2.2 The Ise Shrine exercise

In Ise City located in the Kii Peninsula, Japan, there is *Jingu*, the Ise Shinto Shrine consists of a group of 125 shrines. The central ones are the Inner Shrine (Naiku) dedicated to Amaterasu-Oomikami, the goddess of the sun, founded about 2000 years ago and the Outer Shrine (Geku) dedicated to Toyouke-Oomikami, the goddess of life, about 1500 years ago. The shrine has a unique tradition called *Shikinen Sengu*, to rebuild itself every 20 years that has been exercised since about 1300 years ago (recorded as 690AD) to move goddesses to new shrines. The rebuilding targets are the Inner and Outer shrines, 14 separate (別), 43 related (摂), 24 end (末) and 42 administered

(所管) shrines, other structures including gates and fences, and 1576 pieces of 714 kinds of clothes and treasures. It is not just a building renewal but remake of the whole set of cultural items. Such rebuilding had experienced a number of socio-economic difficulties including over 120 years break during the civil war period in 15–16 centuries but has been continued till now and the 62nd rebuilding was completed in 2013 which started in 2005 (Ise Chamber of Commerce, 2006).

This tradition is quite unique in various respects such as:

(1) Egyptian pyramids or Greek shrines are stone made and last thousands of years but in Japan shrines are made of wood in an old fashion of no foundation stones which cannot last long by simple preservation but can last forever by rebuilding identical under the same design similar to life reproduction.

(2) As long as rebuilding is possible in absolutely the same way, the environmental conditions can be considered sustained. Such rebuilding or renewal of goods becomes impossible if the same wood cannot be obtained for buildings or the same bird feather is not obtainable for a ceremonial decoration. It is also impossible if a society including architects, engineers, supporters of economy, ceremonies and culture who keep the tradition on going is destroyed. It is a sensor of sustainability of nature as well as society. For instance, a bird *Toki* (Nipponia Nippon) was once extinct around 2000 in Japan and a treasure sword with a handle decorated by its feather became unable to be reproduced. But fortunately the same species was in China and kindly donated for artificial breeding to Sado Island, Niigata Prefecture. Now the Shrine gets the feather from a zoo in Ishikawa Prefecture and the tradition has been continuing.

(3) It is by return a means of keeping community united for daily life and any internal or external threats. Rebuilding the shrine is a common objective for the community to unite and work together. There are many steps to go through to rebuild shrines and conduct the due ceremonies. Such include maintaining forests, cutting and planting trees, transferring them to the shrine site, reusing old woods, keeping design blueprints, training architects, carpenters and craftsmen, etc. and a number of ceremonies and festivals for all steps to duly proceed. At each step, communities allocate roles for members and people work together to achieve their assigned mission. Through such occasions, community members get familiar with, know each other and enjoy togetherness on their traditional mission and roles. Figure 6.10 shows an in-river transport of a Japanese cypress tree, an example of many tasks that community members share for reconstruction of Shrine (top) and ICHARM trainees in front of the renewed Gate (bottom).

Figure 6.10 (Top) "*Okihiki*" queue of *Jingu* parishioners to manually pull a new holy timber in the Isuzu River to be used for *Shikinen Sengu* as a main column of the new Shrine. Participants are the members assigned to this particular section of the river in the whole "*Okihiki*" task in land and river (Photo: Courtesy of Ise Jingu). (Bottom) ICHARM Hazard Mapping trainees in front of the Uji Bridge crossing the Isuzu River. (Photo taken by TANAKA Shigenobu on 13 November 2008)

The 20-year cycle is considered reasonable for various reasons. Within 20 years, it was considered that 10 years are for preparation and 8 years for construction. The traditional ceremonies and engineering technologies are difficult to be succeeded if it is longer than 20 years considering life expectancy, necessary learning time, etc. Over 10,000 Japanese cypress trees are necessary for each rebuilding. It used to be supplied from three nearby mountains but extended to remote forests as more than 100 years old trees are necessary. Recent more than 300 years, trees have been supplied from the Kiso mountains in Central Japan. This is from financial reasons, too which is also critical (Ise Chamber of Commerce, 2006).

By keeping such traditions, Ise communities continuously experience collaboration and mutual help, and the sense of unity for a common objective has been developed, kept alive and motivated. As a result, it was demonstrated that at a historical flood so-called "Tanabata heavy rain" on 7 July 1974 by the Baiu front stimulated by Typhoon 8, the Seta River overflew and the 1400 ha of Ise City was inundated but Ise City had none dead while nation-wise dead and missing were 108. Although there was no direct connection identified between disaster preparedness and reconstruction exercises of the Shrine, this success was believed owing to Ise communities' system of mutual help fostered by the custom. Social capital for emergency response is not necessarily developed by drills for disaster preparedness, but rather through daily community activities that people are engaged together.

6.3 TRANSDISCIPLINARY APPROACH

6.3.1 Need of transdisciplinary approach

As repeatedly seen in previous chapters, an integrated approach is indispensable to achieve disaster risk reduction and building resilient society. It is in fact necessary for any other global sustainability issues that contemporary society faces such as climate change, population growth, urban concentration, decline of biodiversity, poverty, conflicts, environmental degradation, etc. They cannot be solved by traditional disciplinary or sectoral approach but an integrated approach is necessary in any nation.

How is an integrated approach realized? It is simply that all related stakeholders work together to achieve a common goal, that is, a transdisciplinary approach.

The United Nations was clearly aware of such needs when they set up MDGs in 2000 and SDGs in 2015. As seen in Chapter 3, the MDG 8th goal and the SDG 17th goal are the partnership in the societal system of finance, science and technology, capacity-building, trade and systemic issues including policy and institutional coherence, multi-stakeholders' collaboration, and data, monitoring and accountability.

In academia, ICSU, ISSC and UNISDR started a program named Integrated Research on Disaster Risk (IRDR) in 2008, and ICSU, ISSC, UNESCO, UNU, UNEP and WMO started the Future Earth program by merging WCRP, IGBP, DIVERSITAS and IHDP in 2014. Both programs called for integration of natural and social sciences, engineering and humanities and collaboration with decision-makers and practitioners. Such programs necessitated the historical merger of ICSU and ISSC into International Science Council (ISC) that was realized on 4 July 2018 (ISC: website on Brief History).

The IRDR science plan (ICSU, 2008) posed a science question "Why, despite advances in the natural and social science of hazards and disasters, do losses continue to increase?" and proposed "integration" of all related disciplines as a strategy to make research outcomes used and curb disaster losses.

The key practice of integration is not only academia with natural and social sciences and humanities but also policy-makers, practitioners and other stakeholders in public, private and civil sectors who are not directly involved in research to work together and put research solutions to practice. True integration is only possible in the stage of implementation.

The same line of thoughts has been again declared in Sendai Framework for Disaster Risk Reduction 2015–2030 in 2015 (UNISDR, 2015) that states in Para 7 of Section I:

> There is a need for the public and private sectors and civil society organizations, as well as academia and scientific and research institutions, to work more closely together and to create opportunities for collaboration, and for businesses to integrate disaster risk into their management practices.

Therefore, the need of transdisciplinary approach is a common agreement at the global level.

6.3.2 Transdisciplinary approach (TDA) for scientific decision-making

First, we elaborate a definition of transdisciplinary approach and necessary conditions for its implementation in practice.

Definition of Transdisciplinary Approach (TDA)

> Transdisciplinary approach (TDA) is an implementation strategy for scientific knowledge-based decision-making to achieve common societal goals by all related stakeholders of all disciplines (natural, social and humanity sciences) and sectors (public, private, academia and civil) working together, going beyond the limit of disciplinary knowledge and sectoral capacities by creating innovative means that make holistic and transformative solutions possible.

In other words, TDA has the following characteristics:

1. There are common societal goals for multiple stakeholders to achieve.
2. To achieve the goals, actions should be determined through scientific knowledge-based decision-making.
3. All related stakeholders participate and work together not only to find research solutions but also to implement them.
4. Solutions go beyond the limit that a discipline or a sector can go alone.
5. Research process does not stop at solution finding but continuously extends to implementation.

Here common societal goals such as various sustainable issues are shared by many people in a society that are too complex to be solved by a disciplinary or a sectoral approach. Research is an indispensable key step to find a solution for a complex problem, but the process should not stop where a solution was found but definitely continue into implementation. The approach ends when a solution is delivered to society and a goal is achieved. The related stakeholders include all disciplinary players of natural and social scientists, engineers and humanities and sectoral stakeholders of public and private sectors and civil society. This makes it possible to achieve something beyond the disciplinary or sectoral capacities.

Thus TDA is different from a narrow sense of "transdisciplinary research" in which finding a solution is the goal to be achieved and assessed, and realization of the solution in society is another matter. In TDA, research is an integral part of the implementation and cannot be separated. It implies that the participants of TDA are, researchers or students, necessarily involved in final implementation process whatever the mode of involvement is. This is a difficult condition for TDA but very important because the real integration of disciplines and sectors can be realized only in an implementation stage. In fact, there are quite a few examples in an implementation stage where TDA was realized with integration of many stakeholders and necessary knowledge and resources for a project. It is because, at the final implementation stage, the project is everybody's business and all stakeholders are willing to participate.

What can TDA do?

To implement a societal project, there are many different ways. TDA is one of them and traditional disciplinary and sectoral approach is more common and popular. But at least to solve problems as complexed as sustainability issues like disaster risk reduction, TDA is indispensable. By TDA the following are possible:

1. Synergy will be created to make all bits of knowledge, resources and capacities available to put to work.

2. Decision-making process will become transparent, accountable and free from corruption.
3. Scientific knowledge-based decision-making can become the mainstream.

Without creating synergy by gathering all potential capacities and going beyond the limits of disciplinary or sectoral approach, such complex societal problems can never be solved to reach a goal. In many cases of sustainability issues, bits of knowledge or experiences are available, but their integration is not realized. TDA can bring a breakthrough by providing a framework for integrating all potential capacities and resources, especially scientific knowledge, into synergy.

Another is that under the presence of all related stakeholders, the decision-making process necessarily becomes "transparent" and can be free from a dictatorship by a certain individual or a certain political or pressure group. The most important aspect of TDA is this transparency in decision-making which is indispensable for many stakeholders to work together. Under a mono-disciplinary or mono-sectoral environment, the traditional know-how or experiences would prevail in the discussion and fresh ideas of laymen may be laughed out. As a result, nobody may voice out "the King is naked!". Furthermore, decision-making in a closed chamber often leads to under-table negotiations and "corruption", which is the most harmful decision-making process in any society.

In addition, the most decisive effect of transparency in decision-making is that scientific knowledge can become central to lead discussion and gain the majority. Then research outcomes can be easily integrated with implementation. Scientific evidence is the clearest and most explainable logic to persuade people and leads scientific knowledge-based decision-making realized. In fact, scientific knowledge-based logic is often not possible if a decision-making process is not transparent. It is easily disturbed by big voices, administrative convenience, political power balance and other irrational logics. Fairness, accountability and scientific knowledge-based decision-making are only possible under a transparent decision-making environment. And it is obvious that contemporary global sustainability issues can never be solved without a full support of scientific knowledge available.

Scientific knowledge-based decision-making

Nevertheless, the transparent decision-making process and scientific knowledge-based decision-making are not popular exercises in any nation. Scientific knowledge-based decision-making should have the following components:

1. Scientific knowledge is centrally and systematically used in the decision-making process: designing, assessing and selecting options.

2. Scientific knowledge of high quality is institutionally made available to decision-making process.

In scientific knowledge-based decision-making, scientific knowledge, instead of other driving forces, should be centrally and systematically used as a basis of decision-making and such a process cannot be realized just by chance but a system to access scientific knowledge should be institutionally installed in an administrative scheme. Political leaders are well aware of the decisive importance of scientific knowledge and try to get the right knowledge as much as possible but their wish is often sealed by their political ambition to get firm support from their constituency or supporting group. As a result, transparency is often put to a minimum if not neglected or becomes merely a political gesture and tends to rely on opinions of friends or arbitrarily selected people. In fact the selection of advisors is a political process and not necessarily to listen to the best available knowledge in scientific community. Very unfortunately, it is not exceptional that selection of alternative actions is made considering political or even personal benefits offered by return, which are the most common sources of corruption.

In order for national or local governments to acquire proper scientific knowledge for decision-making, there are some typical ways such as:

(1) to keep high-level governmental research institutes within governments and rely on their accumulated experiences and judgments,
(2) out-sourcing rather than in-house to rely on high level private or international sector's knowledge of consulting firms, advanced companies or international funding agencies, and
(3) to create a governmental consultation scheme to utilize the advanced knowledge of local universities, public and private institutes and many experienced experts.

Especially the third type is widely exercised in Japan, that is, governments often create a consultation committee, commission or council consisting of multi-disciplinary and multi-sectoral members and ask them to investigate, discuss and deliver recommendations on requested matters when difficult societal decisions should be made. If the science sector is well represented in such consultation scheme and final decisions well appreciate their recommendations, scientific knowledge-based decision-making will be realized to a good extent.

Knowledge flow infrastructure

In such a scheme of feeding the best available knowledge into decision-making at various levels is of vital importance to keep the best use of scientific knowledge. But whichever process the government takes, the representatives of science sector should be well equipped with the best available knowledge. In an ideal case, either in-house institutes, local universities, or competent companies have experienced experts with top scientific knowledge. But most often it is not necessarily the case and even a distinguished expert is available, he/she can cover its special area and not all. This is why knowledge flow infrastructure is necessary to be established and installed available in high quality at any level of decision-making. Such infrastructure would be a network of universities, research institutes, professional associations, companies' training mechanism, recurrent learning system, etc. where knowledge should be institutionally kept smoothly flowing in the network. If such a network is not institutionally supported by an official scheme such as governments, the assigned nodes of knowledge providers cannot properly function. The knowledge flow infrastructure may thus be characterized as follows:

1. Knowledge flow is a network to connect scientific knowledge sources with users and to make necessary and best practicable knowledge accessible on any occasions of the decision-making process in society.
2. Knowledge flow network is ideally established, in function, in parallel to administrative decision-making network to make sure proper scientific knowledge is available whenever needed in decision-making.
3. Such knowledge flow should be supported by an institutional mechanism to be continuously updated and quality controlled and, with training mechanism, make science interpreters and guides available for decision-makers.

The third condition, availability of science interpreters would be particularly important to make knowledge flow to function. Science sector should produce and maintain such professionals who devote themselves to deliver situation-specific practical knowledge with its applicability, limitation, life cycle effects, etc. As most societal problems need a solution unique to each case under different circumstances and a choice depending on its societal value system, there is no way to replace a transdisciplinary scientific knowledge-based decision-making process by an artificial intelligence (AI). It can only be a decision support system. Decision itself should be made under the transdisciplinary approach supported by a sustainable human empowerment mechanism.

6.3.3 Some case studies

Along this line of thoughts, the Asian Civil Engineering Coordinating Council (ACECC) Technical Committee 21 (TC21) "Transdisciplinary approach for building societal resilience to disasters" was established in 2016 (ACECC: website on TC21). Figure 6.11 shows the aim of TC21, that is, to promote

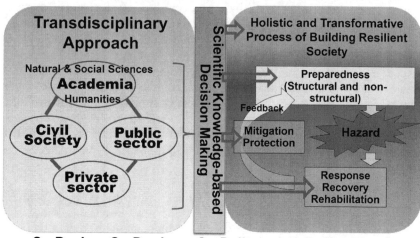

Figure 6.11 The aim of ACCEC TC21 to promote mitigation of disasters by introducing scientific knowledge-based decision-making under TDA. (ACECC: website on TC21)

mitigation of disasters triggered by natural hazards by introducing scientific knowledge-based decision-making through TDA.

In order to achieve this aim, TC21 has been conducting case studies in some nations to find out the reality of the decision-making process in the area of disaster management. To follow are a brief summary of some findings obtained so far in Japan, the Philippines and Nepal. More reports are available at (TC21-JSCE Editorial Group, 2019).

Japanese case

As a method of scientific decision-making, the use of consultation committees is typically exercised in Japan. The Central Disaster Management Council (CDMC) was established in Cabinet Office in 1961 under the Basic Law of Counter Measures for Disasters (BLCMD) (e-gov: website on BLCMD) is now serving as a national platform for disaster risk reduction in Japan which corresponds to the Guidelines: National Platforms for Disaster Risk Reduction (UNISDR, 2007). The guideline says:

> To address the complexity of DRR, Member States of the UN Economic and Social Council, through its Resolution 1999/63, called on

all Governments to maintain and strengthen the established national multi-sectoral platforms for disaster reduction in order to achieve sustainable development goals and objectives, with the full utilization of scientific and technical means.

Figure 6.12 shows the formation of the Central Disaster Management Council (CDMC) of the Cabinet Office (CO), Japan which is made up of all Cabinet members and representatives of the Bank of Japan, Japanese Red Cross Society, NHK, NTT plus several experienced experts including one from ICHARM, Public Works Research Institute (PWRI) (CO: website on CDMC, Membership). The standing membership of NHK (Nippon Hoso Kyokai or Japan Broadcasting Corporation) in this Council is the basis of its outstanding role in warning and emergency instruction against hazards and disasters. Cabinet office home page explains the detailed structure, that is, under the council, there are three special investigation committees for implementation of disaster risk reduction, investigation of methodologies and specific to the Tohoku tsunami (CO: website on CDMC, Organization). All those special committees have a number of experts from academia as well as public, private and civil sectors. Besides, there are a number of working groups under the CDMC or its special committees including ones

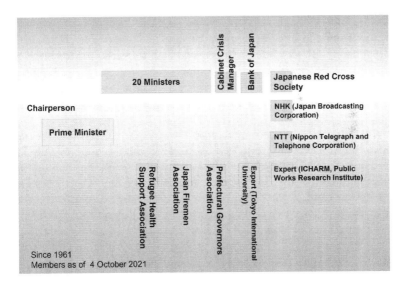

Figure 6.12 The formation of the Central Disaster Management Council (CDMC) of the Cabinet Office (CO), Japan (CO: website on CDMC, Membership). It serves as the Japan National Platform for DRR, well observing a guideline for a multi-sectoral platform. Scientific knowledge-based decision-making seems rather well supported by the institutional framework.

on large scale evacuation from flood and storm surge, and others on preparedness for the anticipated Nankai (Southern Sea) Trough Earthquake. Those working groups are usually chaired by a university professor and members are dominantly experts from academia and responsible officers in the public sector.

Besides, in order to promote awareness and disaster consciousness of the people, the National Congress of Promoting Disaster Prevention (CO: website on CDMC, National Congress) was founded in 2015 consisting of leaders of 40 associations (as of 2015) of such as primary to higher-level schools, local governments, mayors, women, medical doctors, nurses, journalists, firemen, laborers, handicapped, consumers corporation, etc. and holds large annual conventions to exchange ideas of self-help and mutual-help experiences.

The Philippines case

(1) Ormoc City

A devastating flood by Typhoon Thelma (local name Uring) hit Ormoc City, Leyte Island, the Philippines on 5 November 1991. Dead and missing were about 8000 mainly along the Anilao and the Malbasag Rivers in the city. It was the biggest flood disaster experienced in Ormoc, corresponding to the scale of once in 50 years. In recovery, JICA constructed three slit dams and did channel improvements to prevent such tragic floods to repeat again (JICA, 2014), that were tested by similar Typhoon Koni in 2003 and were demonstrated successful. Maintenance work for the infrastructure has been done by transdisciplinary approach involving all stakeholders, national, regional and communities. It solved the issue of informal residents too, simply by constructing lining fences on the river dikes. The construction of simple fence worked out very well to keep potential squatters out. This is not because of the fence itself but because all the people in the community observed and cooperated for the implementation of the policy. It was maintained by the Ormoc City Flood Mitigation Committee consisting of central and city governments, and barangays, the smallest administrative unit in the Philippines. River maintenance works are done by communities. Fence became the solution to the squatters' problem by transdisciplinary approach, which would be an extremely rare and valuable example of community practices in the world.

(2) Tacloban City

An unprecedented strong typhoon Haiyan (local name Yolanda) hit Tacloban City, Leyte Island and brought tsunami-like storm surges on 8 November 2013. The number of people killed was 6300, missing 1602, damaged houses 1.14 million and people served inside or outside of evacuation centers

5.1 million (Philippines NDRRMC: website on final report on Yolanda). About the dead and missing, Yamada (2017) reported that including families who had no survivors to claim, the estimate ranged up to 18,000. President Aquino III committed building back better which was followed by declaration of "40 m no-build zone" policy by the Department of Environment and Natural Resources (DENR) and the Department of Public Works and Highways (DPWH) put up its signs in the coastal zones in February 2014 (Yap, 2014; Yamada, 2017).

This policy resulted a very quick recovery as construction restriction was only for the narrow 40 m no-build zone from the coastline allowing the rest of the area free to restore houses to the original state of order. But still in the 40 m no-build zone and additional areas occupied by illegal residents, there were about 14,600 houses to be replaced in nearby areas. In order to decide on the replacement sites and necessary supports, an intensive consensus-building process took place in a transdisciplinary manner involving 138 barangay chiefs, local government officers and community people. They agreed and cooperated to build new residential areas in the northern Tacloban district including commercial shops, schools, etc. Many people already moved in by 2017 although the residents had various difficulties such as lack of water supply (Takeuchi and Tanaka, 2016; Iuchi and Maly, 2016; Tuhkanen et al., 2018).

Nepal case

The Gorkha earthquake of 7.8 Mw and max Mercalli Intensity IX (Violent) occurred in central Nepal at 11:56 NST on 25 April 2015 which killed nearly 9000 people and injured more than 22,000 causing $7 Billion (about one-third of FY2013–2014 Nepal GDP) economic losses. About 500,000 houses were fully damaged and 250,000 partially damaged. The 95% of fully damaged houses were simple masonry houses pasted by mud and no cement nor any other reinforcement even by bamboo (National Planning Commission, 2015).

The reconstruction process took place by near transdisciplinary approach including various public sectors and communities. The damaged brick houses were decided to recover by bricks again but with the *masonry band method* which strengthen the brick structure by introducing band frames in plinth, lintel, roof, doors, windows, etc. as seen in the top of Figure 6.13. The unique transdisciplinary approach was in its implementation, that is, the Department of Urban Development and Building Construction (DUDBC) and the National Reconstruction Authority (NRA) decided on the use of a "sill and lintel band method" for masonry buildings and the Finance Ministry supported this method by giving residents 300K Nepal Rupee Grants subsidy who followed the method for rebuilding their houses. With agreement of the method 50K Nepal Rupee, when plinth was completed by the method 150K Rupee and when roof (band) was completed 100K Rupee were given to the house owner. Another important part of TDA here was technological community, who trained about 3200 engineers recruited from the government for them to

Figure 6.13 A reconstruction site in Chautara, about 40 km east of Kathmandu, where reconstruction of masonry house in band method is taking place and local residents are watching at in front of a new house built earlier. (Photo taken by the author on 21 April 2017)

be able to guide, inspect and judge the *masonry band method* and its completion stage in house construction (Paudyal, 2019). This whole process had been participated and supported by local communities such as deciding the order of construction according to the urgency of the family's need of a house.

Preliminary findings

Although the cases are limited, some important findings on TDA were found which might be summarized as follows:

(1) Scientific knowledge-based decision-making is a difficult strategy to be implemented especially in TDA manner. But there are some good practices available that would help learning each other.
(2) Even though the selected science and technology may be reasonable, its decision-making process is often top-down and not necessarily clear such as on involved players, assessment and selection processes, etc.
(3) The main players are mostly the public sector and local communities, and roles of private sector including media, construction companies and academia such as local universities are unclear.
(4) TDA is often well-practiced in a local level in a project implementation stage rather than a high-level decision-making stage.

In Japan, transdisciplinary councils and committees are formed to support governmental decision-making in national as well as prefectural and city levels. Thus at least from the structural point, the transdisciplinary approach for scientific-knowledge-based decision-making is well institutionalized. But from the point of the supporting network of the knowledge flow mentioned earlier, there are still a number of challenges left especially in the local levels. The situation in other nations seems, however, quite different and there are many barriers for the TDA concept to be implemented in societal decision-making. Looking at decision-making process in society itself would be the starting point. The efforts of advocating and exchanging experiences of TDA for scientific knowledge-based decision-making would be important.

REFERENCES

ACECC: website on TC21. Transdisciplinary Approach (TDA) for Building Societal Resilience to Disasters. www.acecc-world.org/TC/TC21/index.htm (accessed 28 September 2020).

CO (2020) *On the Damages by 2019 Typhoon 19 and Others*. Cabinet Office, Report on 10 April 2020. www.bousai.go.jp/updates/r1typhoon19/pdf/r1typhoon19_45.pdf (accessed 21 December 2020).

CO: website on CDMC, National Congress. Central Disaster Management Council, Cabinet Office. www.bousai.go.jp/kaigirep/suishin/ (accessed 28 September 2020).

CO: website on CDMC, Membership. Central Disaster Management Council, Cabinet Office. www.bousai.go.jp/kaigirep/chuobou/pdf/meibo.pdf (accessed 1 November 2021).

CO: website on CDMC, Organization. Central Disaster Management Council, Cabinet Office. www.bousai.go.jp/kaigirep/index.html (accessed 28 September 2020).

CO: website on Flood Fighting Teams. www.bousai.go.jp/kaigirep/hakusho/h14/bousai2002/html/honmon/hm130202.htm (accessed 4 December 2021).

CO: website on Nagasaki Water Disaster. www.bousai.go.jp/kaigirep/houko kusho/hukkousesaku/saigaitaiou/output_html_1/case198201.html (accessed 21 December 2020).

e-gov: website on BLCMD. Basic Law of Counter Measures for Disaster. https://elaws.e-gov.go.jp/document?lawid=336AC0000000223 (accessed 28 September 2020).

FDMA (2019, January 9) *Damages and Responses Situations of 2018 July Heavy Rains and Typhoon 12.* (Report no. 59th) at 13:00 on 9 January 2018, Japan Fire and Disaster Management Agency (FDMA), Emergency Response Office. h30–7_59.pdf (fdma.go.jp) (accessed 21 December 2020).

Hiroshima City: website on the Outline of the July 2018 Heavy Rain Disaster. www.city.hiroshima.lg.jp/uploaded/attachment/54798.pdf (accessed 21 December 2020).

Hiroshima Prefecture: website on 20 August 2014 Heavy Rain Disaster. www.city.hiroshima.lg.jp/soshiki/126/5585.html#pagetop (accessed 19 December 2020).

ICSU (2008) *A Science Plan for Integrated Research on Disaster Risk.* www.irdrinternational.org/wp-content/uploads/2011/06/IRDR%20Science%20Plan.pdf (accessed 8 January 2021).

IPSS (2012) *Future Projection of Japanese Population. 2012 Projection.* National Institute of Population and Social Security Research, Population Research Series. www.ipss.go.jp/syoushika/tohkei/newest04/gh2401.pdf (accessed 6 January 2021).

IPSS (2017) *Future Projection of Japanese Population. 2017 Projection.* National Institute of Population and Social Security Research, Population Research Series No. 336, 31 July 2017. www.ipss.go.jp/pp-zenkoku/j/zenkoku2017/pp29_ReportALL.pdf (accessed 6 January 2021).

ISC: website on Brief History. https://council.science/about-us/a-brief-history/ (accessed 8 January 2021).

Ise Chamber of Commerce (2006) *Certificate Exam: O-Ise-san.* Official Guidebook, 24 April 2006.

Iuchi, K. and E. Maly (2016) Recovery Plans and Progress. In: IRIDeS (2016) *Typhoon Haiyan: Report on the Ongoing Process of Building Back Better,* Chapter 2, Tohoku University, Sendai, July 2016.

Japanese Law No. 22 (2002) *Urban Revitalization Special Law.* https://elaws.e-gov.go.jp/document?lawid=414AC0000000022 (accessed 5 January 2021).

Japanese Law No. 39 (2014). www.mlit.go.jp/report/press/toshi07_hh_000079.html (accessed 12 October 2021).

Japanese Law No. 100 (1968, June 15) *City Planning Law.* www.japaneselawtranslation.go.jp/law/detail/?ft=1&re=01&dn=1&ia=03&ja=04&x=56&y=10&co=01&ky=city+planning&page=2 (accessed 12 October 2021).

JICA (2014, April) Support for Typhoon-Stricken Leyte Island. *JICA's World*, 5(5), 4–5. www.jica.go.jp/english/publications/j-world/c8h0vm00008qskjz-att/1404. pdf (accessed 10 January 2021).

JMA (2014) *August 2014 Heavy Rains*. Fast Report at Disaster. Natural Phenomena Report of Disaster. No. 4, 17 November 2014. www.jma.go.jp/jma/kishou/ books/saigaiji/saigaiji_201404.pdf (accessed 24 November 2021).

JMA (2018, July 13) *July 2018 Heavy Rains by the Frontal Activities and Typhoon 7. Meteorological Cases Resulted in Disasters*. Japan Meteorological Agency. www.data.jma.go.jp/obd/stats/data/bosai/report/2018/20180713/jyun_ sokuji20180628-0708.pdf (accessed 20 August 2020).

JMA (2019, October 15) *Heavy Rains and Storms by Typhoon 19. Meteorological Cases Resulted in Disasters*. Japan Meteorological Agency. www.data.jma. go.jp/obd/stats/data/bosai/report/2019/20191012/jyun_sokuji20191010-1013. pdf (accessed 22 December 2020).

JMA: website on Changes in Extremes. www.data.jma.go.jp/cpdinfo/extreme/ extreme_p.html (accessed 1 October 2020).

JMA: website on Historical Ranking of Extreme Records. www.data.jma.go.jp/obd/ stats/etrn/view/rankall.php (accessed 29 September 2020).

JMA: website on JMA History. www.jma.go.jp/jma/kishou/intro/gyomu/index2. html (accessed 8 April 2021).

JMA: website on Typhoon Statistics. www.data.jma.go.jp/yoho/typhoon/statistics/ index.html (accessed 16 March 2022).

JSCE (2020) *Investigation Report on Heavy Rain Disasters by 2019 Typhoon 19*. Committee on Hydroscience and Hydraulic Engineering, Japan Society of Civil Engineers. https://committees.jsce.or.jp/report/system/files/taifu19.pdf (accessed 29 April 2021).

Kato, Teruyuki (2020) Quasi-stationary Band-shaped Precipitation Systems, Named "Senjo Kousuitai", Causing Localized Heavy Rainfall in Japan. *Journal of Meteorological Society of Japan*, Ser. II. 98(3), 485–509. https://doi.org/10.2151/ jmsj.2020-029.

MIC (2021) *Population Estimate, January 2021*. Statistics Bureau, Ministry of Internal Affairs and Communication. www.stat.go.jp/data/jinsui/pdf/202101.pdf (accessed 6 May 2021).

MLIT (2008) *Climate Change Adaptation Strategies to Cope with Water-related Disasters Due to Global Warming*. Recommendation of River Sector Committee of Panel on Infrastructure Development Subcommittee on Climate Change Adaptation for Flood Control, November 2007. www.mlit.go.jp/river/ basic_info/jigyo_keikaku/gaiyou/kikouhendou/pdf/interimreport.pdf Reference. www.mlit.go.jp/river/basic_info/jigyo_keikaku/gaiyou/kikouhendou/pdf/draft policyreportref.pdf (accessed 24 November 2021).

MLIT (2015, March) *Towards Formation of Compact Cities*. The Compact City Formation Supporting Team. www.mlit.go.jp/toshi/city_plan/toshi01_sg_000237. html 1st meeting. Reference no. 3. www.mlit.go.jp/common/001083358.pdf (accessed 29 December 2020).

MLIT (2018) *Investigation Committee on Promoting Flood Fighting Team Activities*. 1st Meeting on 19 March 2018. www.mlit.go.jp/river/shinngikai_blog/sui boukatsudou_kasseika/dai01kai/ Reference 2: Real State of Flood Fighting Teams. www.mlit.go.jp/river/shinngikai_blog/suiboukatsudou_kasseika/dai01kai/ dai01kai_siryou2.pdf (accessed 4 December 2021).

MLIT (2019a, November 22) *Damages by 2019 Typhoon 19*. River Sub-Committee, MLIT Council of Social Capital. Material no. 6. www.mlit.go.jp/river/shinngikai_blog/shaseishin/kasenbunkakai/shouiinkai/kikouhendou_suigai/1/pdf/11_R1T19niyoruhigai.pdf (accessed 23 December 2020).

MLIT (2019b) *001326032.pdf* (mlit.go.jp) (accessed 11 October 2021).

MLIT: website on the Abukuma River. www.thr.mlit.go.jp/bumon/b00037/k00290/river-hp/kasen/outline/kokudo/kasen/0012.html (accessed 22 December 2020).

MLIT: website on Compact City Promoting Project. www.mlit.go.jp/en/toshi/city_plan/content/001422663.pdf. Budgets of FY 2021 for Promoting Compact City. www.mlit.go.jp/toshi/city_plan/toshi_city_plan_tk_000032.html (accessed 1 December 2021).

MLIT Numazu River and Road Office (2019, October 18) *The Kano River Discharge Report, No. 2.* www.cbr.mlit.go.jp/kawatomizu/shussuijyoukyou/pdf/indexR01_taihu19/kanogawa-2.pdf (accessed 22 December 2020).

MLIT Urban Bureau (2017, February 15) *Current Status and Agenda on Urban Planning*. Sub-committee on Urban Planning Basic Problems. www.mlit.go.jp/policy/shingikai/s204_toshikeikakukihonmondai_past.html 1st Meeting, Reference No. 3. www.mlit.go.jp/common/001180304.pdf (accessed 5 January 2021).

Nagano City (2021) 2019 *East Japan Typhoon: Nagano City Disaster Report*. Nagano City, March 2021. https://www.city.nagano.nagano.jp/uploaded/attachment/363607.pdf (accessed 12 August 2022).

National Planning Commission (2015) *Nepal Earthquake 2015 Post Disaster Needs Assessment. Vol. A: Key Findings*. Government of Nepal. www.worldbank.org/content/dam/Worldbank/document/SAR/nepal/PDNA%20Volume%20A%20Final.pdf (accessed 29 November 2021).

OECD (2012) *Compact City Policies: A Comparative Assessment*. www.oecd.org/regional/greening-cities-regions/compact-city.htm (accessed 29 December 2020).

Okayama Prefecture: website on the Post Audit Report on the July 2018 Heavy Rain Disaster. www.pref.okayama.jp/uploaded/life/601705_5031910_misc.pdf (accessed 21 December 2020).

Paudyal, Youb Raj (2019) Institutional Structure for Incorporating Science into Decision-making Reconstruction Process and Practices after the Gorkha Earthquake, 25 April 2015. In: TC21-JSCE Editorial Group (ed.), *Transdisciplinary Approach (TDA) for Building Societal Resilience to Disasters-Concepts and Case Study for Practicing TDA*, Chapter 8, 81–86. Published by JSCE, September 2019. www.acecc-world.org/TC21/index.htm (accessed 29 November 2021).

Philippines NDRRMC: website on Final Report on Yolanda. https://ndrrmc.gov.ph/attachments/article/1329/FINAL_REPORT_re_Effects_of_Typhoon_YOLANDA_HAIYAN_06-09NOV2013.pdf (accessed 13 August 2022).

Study Team for In-town Residence (2000) *Towards Construction of Compact City*. MLIT Tohoku Regional Bureau. www.thr.mlit.go.jp/compact-city/contents/machinakakyozyu/index.html (accessed 7 January 2021).

Takeuchi, K. and S. Tanaka (2016) Recovery from Catastrophe and Building Back Better. *JDR*, 11(6), 1190–1201. http://doi.org/10.20965/jdr.2016.p1190.

TC21-JSCE Editorial Group (2019) *Transdisciplinary Approach (TDA) for Building Societal Resilience to Disasters -Concepts and case study for practicing TDA-.*

JSCE, Tokyo, 2019. ISBN: 978-4-8106-1014-7. https://tc.acecc-world.org/21/ (accessed 18 July 2022).

Tuhkanen, Heidi, M. Boyland, G. Han, A. Patel, K. Johnson, A. Rosemarin and L.L. Mangada (2018) A Typology Framework for Trade-offs in Development and Disaster Risk Reduction: A Case Study of Typhoon Haiyan Recovery in Tacloban, Philippines. *Sustainability*, 10(6). http://doi.org/10.3390/su10061924.

UN DESA Population Division (2017) *World Population Prospects: The 2017 Revision, Volume I: Comprehensive Tables (ST/ESA/SER.A/399)*. https://population.un.org/wpp/Publications/Files/WPP2017_Volume-I_Comprehensive-Tables.pdf (accessed 24 December 2020).

UN DESA Population Division (2019) *World Urbanization Prospects 2018: Highlights (ST/ESA/SER.A/421)*. https://population.un.org/wup/Publications/Files/WUP2018-Highlights.pdf (accessed 24 December 2020).

UNISDR (2007) *Guidelines: National Platforms for Disaster Risk Reduction*. www.preventionweb.net/files/601_engguidelinesnpdrr.pdf (accessed 25 November 2021).

UNISDR (2015) *Sendai Framework for Disaster Risk Reduction 2015–2030*. www.preventionweb.net/files/43291_sendaiframeworkfordrren.pdf (accessed 8 January 2021).

Weathernews: website on Typhoon 19 is Greater Than Typhoon 15. https://weathernews.jp/s/topics/201910/080085/ (accessed 30 September 2010).

Yamada, Seiji (2017) Disaster Capitalism in Post-Yolanda/Haiyan Eastern Visayas. *Environmental Justice*, 11(2). http://doi.org/10.1089/env.2017.0030.

Yap, D.J. (2014) Group Denounces "No-Build Zone" Policy. *Philippine Daily Inquirer*, 04:49 pm, 4 April 2014. https://newsinfo.inquirer.net/591772/group-denounces-no-build-zone-policy#ixzz6uQMYfpAH (accessed 10 May 2021).

Epilogue

The deadliest disasters that human beings ever experienced were famines. Millions of people died or migrated, cultivated lands were abandoned and villages disappeared. Causes of famines range from wars, conflicts to various natural hazards. Floods, droughts, cold weather, prolonged rains, epidemics, pests, volcanic eruptions etc. are all potential causes. But famines are not necessarily a direct result of those hazards but a prolonged effect of the aftermath of a disaster. In case of floods, for example, if only the people drowned in water or killed by debris are counted, the death toll would be much smaller. Reports of death tolls of the 1931 Yangtze-Huai flood in China range from 140,000 to 4 million people in literature (Wikipedia: 1931 China floods). The smallest must be a record of immediate death by drowning or debris flows and the largest a rough estimate after the following famine and epidemics such as 1932 cholera. Thus disaster recovery needs several years and, if not well managed, several decades by falling into a vicious cycle in the poverty-disaster risk nexus (UNISDR, 2009b).

In the Edo Era, a number of famines attacked Japan, of which Kyoho famine in 1732, Tenmei famin (1782–88) and Tenpo famin (1833–39) were serious and often called Three Great Famines of the Edo Era. They made a countless number of people starve to death or flee from villages, and many farmlands were abandoned.

Tenmei Famine (1782–88) was the most serious one mainly hit central to northern Japan, first by cold weather in 1782 followed by eruptions of Mt. Iwaki (1782.11–1783.6) and Mt. Asama (1783.4–7). It is considered that Mt. Asama's eruption was most influential and about 900,000 people died of hunger or epidemics (Wikipedia: Tenmei Famine) and in many ruined villages the impact lasted several decades till Tenpo famine a half-century later.

In the same period, there were huge volcanic eruptions in Iceland, Mt. Laki (1783.6–1784.2) and Mt. Grimsvotn (1783–1785). According to (Wikipedia: Laki), Mt. Laki's eruption was truly serious having

DOI: 10.1201/9781003275541-8

discharged 14 km³ of basalt lava and a huge amount of volcanic ash. Its impact lasted for years with cold weather in the Northern Hemisphere that brought famines in various parts of the world including India and north Africa. In Iceland, it killed 1/4 of population and 50% of live stocks of the nation. It seems that the Tenmei famine was mainly by the effect of Mt. Asama eruption but that of Mt. Laki also influenced and worsened the situation.

Tenpo famine (1833–1839) was due to heavy rain, floods and cold weather and northeastern Japan got serious damages that gave impacts to western Japan, too. Many villages had not recovered from Tenmei famine yet and their population and production had been kept low. By the additional famine, the impact was great. In Edo, the government opened 21 emergency salvation houses with a capacity of 5100 persons per house where 700,000 people were said to have gathered. In Osaka, OSHIO Heihachiro's Rebellion occurred who tried to overturn the Tokugawa administration. In Kai (now Yamanashi Prefecture) country-wide farmers' riots occurred (Wikipedia: Tenpo Famine).

It was only less than a decade later from Tenpo famine that in Ireland a truly devastating famine occurred for 1845–1849. It was due to late blight disease of potato. Ireland then had a monoculture on potato of a mono species to eat and export to Britain which was totally lost for years. This has been called the Great Famine (or Irish Potato Famine), the largest disaster in Europe in the 19th century which not only killed a million people (1/8 of the total population of Ireland) but also gave a great impact to the outer world by sending about two million immigrants to the new world. During the 1840s, about a half of all immigrants to the United States were Irish and by the time when it achieved independence from the UK in 1921, the population was barely half of that in the early 1840s (Mokyr, 2020).

NINOMIYA Kinjiro Sontoku lived exactly in such a period 1787–1856 in Japan and devoted his life for recovery of villages devastated by the Tenmei famine in the 1780s and revitalized them to safely go through Tenpo famine in the 1830s. He was a societal reformer to recover distressed villages by their own empowerment. The philosophy of his recovery method called "Hotoku-shiho" would serve as a universal guide for disaster recovery and resilience building of a society. It is a set of self-reliant, diligent, mutually trusting, mutually helping, bottom-up and top-down integrated, transdisciplinary and economically practical principles as briefly summarized in Figure E.1. Introduction of his life would explain how those principles were developed and suggest what individual disaster managers to keep in mind. He sincerely and eloquently demonstrated that human empowerment is the only sure way for sustainable development.

報徳仕法 "Hotoku-shiho"
Methodology of repaying for the virtue of others

- 至誠 (Shisei) Act sincerely to others.
- 勤労 (Kinro) Work hard to repay for the virtue of others.
- 分度 (Bundo) Keep living standard within your income.
- 推譲 (Suijo) Invest the residuals for the future of your family and others.
- 積小為大 (Sekisho-idai) Accumulate little efforts to a great.
- 一円融合 (Ichien-yugo) All players fused into one. Make every small part work together to achieve an aimed objective.

Figure E.1 Principles of Hotoku-shiho.

NINOMIYA KINJIRO SONTOKU AND HOTOKU-SHIHO

NINOMIYA Kinjiro Sontoku has been popular since the Meiji Era as a symbol of a hard worker and a humanitarian social reform leader. His statues were built in many elementary schools in Japan. After World War II, his portrait was selected for one yen bill issued from 1946 to 1958. In the Meiji Era one of the most respected religious leaders and philosophers UCHIMURA Kanzo wrote about him as one of five representative Japanese in his book "Representative men of Japan" (Uchimura, 1908). In this book, he described Kinjiro as a social reformer believed in "moral power" to recover a village. Uchimura introduced an episode that when the lord of Odawara domain requested Kinjiro to recover the devastated Sakuramachi, he investigated the villages and reported the lord the following:

> The art of love (仁術) alone can restore peace and abundance to those poor people. . . . Grants in money, or release from taxes, will in no way help them in their distress. Indeed, one secret of their salvation lies in withdrawing all monetary help from them. Such help only induces avarice and indolence and is a fruitful source of dissensions among the people. The wilderness must be opened by its own resources, and poverty must be made to rescue itself.

Kinjiro indeed believed in people's mind as the key to recovery. Although this episode might be a truth, and it indeed reflects Kinjiro's spirit, there is no written evidence of this statement by Kinjiro himself but only in his followers' books "Hotokuki" (Tomita, 1883) and "Ninomiyaou-yawa" (Fukuzumi, 1933) with many other legendary episodes. Avoiding unfounded stories, NINOMIYA Yasuhiro, a descendant of the Ninomiya family, wrote a book "Life and thoughts of NINOMIYA Kinjiro" using only the documents that Kinjiro himself wrote. From this, the image of Kinjiro as a practical social reformer, an efficient disaster recovery specialist and a manager of micro-economy becomes clear. To follow are mainly the author's summary and interpretation of this book (Ninomiya, 2008).

Birth of NINOMIYA Kinjiro and tragic flood disasters

NINOMIYA Kinjiro (later he called himself as Sontoku at the age of 56) was born in a middle-class landowner's family as the first son in 1787 near Mt. Fuji in Kayama Village, Ashigarakami-gun (now Odawara City), Sagami (now Kanagawa Prefecture). The village was hit by a flood of the Sakawa River and his family lost their house and paddy fields in 1791 (Kinjiro was 5 years old, counting the birth year as one year old. Kinjiro's age is provided hereafter in parentheses). His father became sick because of this devastation and died in 1800 (14 years old). His mother also died two years later (16 years old). Still worse, in the same year of his mother's death in 1802, a serious flood breached the Sakawa River levee and hit again the Kayama Village and he lost all paddy fields. By this devastation, there was no other choice than to separate the family. His two brothers went to his mother's parents' house and himself to his uncle's house. The Sakawa River had frequent floods after Mt. Fuji eruption in the Hoei Era in 1707 as it deposited a huge amount of sediment and the influence to the Sakawa River lasted a century or so (Sumiya et al., 2002; Ninomiya, 2008).

He worked hard in his uncle's house with creative ideas. He picked up some rice seedlings thrown away and planted them on ridge between rice paddies and harvested a straw bag (60 kg) of rice. He lent the money he got from its sale instead of using it and saved. This experience developed his policy of "sekisho-idai", that is, accumulate a little to a great. He also cultivated rapes to get rapeseed oil by which he could buy books and study them.

Recovery of his own house

In 1805, at the age of 19, Kinjiro returned to his own house to restore it. By his hard work and savings, he could start buying the paddy fields back one by one. In 5 years by 1810 (24 years old), 10th anniversary of his father's death, he could restore the house. The ways of raising restoration funds were such as collecting and selling firewood, lending paddy fields and money, engaging in rice speculation, operating a mutual financing association ("kou" in Japanese) etc. Also he requested contribution from members of the Ninomiya family and gave awards for the members who contributed most. He considered restoration of the head house of a family was for co-prosperity of all family members. He himself kept a very frugal life. He always saved residuals to invest for the future of all family members which became his basic principle "Suijo", i.e., invest residuals for the future. During this time he composed *haiku* poems and got haiku poet name "Sansetsu (mountain snow)". In 1810 he climbed up Mt. Fuji and visited Konpira, Kyoto, Osaka, Nara, Ise Shrine, etc. for 49 days.

In 1811 (25 years old), he left Kayama Village and lived in the capitol of the Odawara domain to work for the Kojima and the Hattori families as a retainer. So he became a land owner engaged in tenant farming, rice and firewood sales, mutual financing, money lending, rice speculation, and in addition, worked at high-ranked official's family for salary. He was a diversified business operator.

Recovery of the Hattori family

Kinjiro's such activities soon became known widely, and in 1811, the Hattori family, a top vassal (Karou) of the Odawara domain, requested him the recovery of their financial situation. Then he strongly requested the head of the family, HATTORI Juroubei, "Bundo", that is, to live within his income. It was so difficult for Juroubei to observe it. This was because the domain leaders could not accept the idea that the safety and happiness of the people bring the safety and happiness of the leaders instead of the other way around. The family had nominal yield 1200 "koku" (1 koku is about 150 kg) of rice but in fact they had only 403 "koku" within which they could not live and the debt increased.

In 1817 (31 years old), he was formally assigned to recover the Hattori family's economy. He once again requested Bundo to the family but they could not observe it, partly because the double household became necessary as Juroubei had to move to Edo as the domain lord OOKUBO Tadazane moved there. He tried to make a profit by rice speculation but this time he failed and lost. But he could borrow a 15-year low-interest loan from the Odawara domain in 1820 (34 years old) and created a system of financing association "Gojoukou" for high-rank officers to pay back the high-interest debt replacing it with a low-interest loan and for low-rank officers, same with no-interest loan. By this new lending system, he could keep running the fund repeatedly and all the debts were cleared except the loan from the domain. He also unified a box scale for tax to measure rice which helped farmers a lot by just taxing. By those the 1st phase of the Hattori family recovery completed. Later, however, since the burden of paying back the loan in 15 years was heavy, it did not continue and by 1831 the debt became large again. Kinjiro finally gave his own fund (later called as "Hotoku-kin") solve the situation in 1838.

Recovery of Sakuramachi Villages

In 1821 (35 years old) Kinjiro was requested by the lord of the Odawara domain OOKUBO Tadazane to recover the territory of Ookubo's branch family, the UTSU Hannosuke in Sakuramachi, Shimotsuke (now in Moka City, Tochigi Prefecture). Three villages of Sakuramachi were so devastated after Tenmei famine in 1782–1788, damaged fatal by the eruption of Mt. Asama in 1783. Many people died, many families left villages, and the abandoned paddy fields or vegetable fields turned into wasteland. Population and number of houses were 722 and 156 in 1821 which were considered about 1/3 of the past. As a result, although the nominal yield of rice of Sakuramachi was 4000 koku but in reality, the tax income was only 1000 bags (1 koku=2.5 bags of which tax was 40%=1 bag).

Based on those findings, he requested the lord of the Odawara domain to accept strong Bundo to make village development possible by Suijo. Because, if the extra bags produced by farmers go to tax, there would be no improvements possible in farmers' life and no investment for the future that would spoil farmers' incentive for hard work. In 1823, he got a promise of Bundo from the lord to keep tax 1005 bags and gold 144 ryo (from vegetable fields) for the head of Sakuramachi territory, UTSU Hannosuke for 10 years. The funds given to Kinjiro were 200 bags and gold 50 ryo per year, about which no reporting was required. Also it was assured that there would be no recall to Odawara for 10 years. He got really a 10-year free hand contract. He already moved to Monoi Village, Sakuramachi in 1822 (36 years old) and his whole family in 1823, having sold all houses and properties in Kayama Village.

The Sakuramachi house he lived in is preserved as it was in the same place in the back yard of the Ninomiya Sontoku Shiryokan Museum as shown in Figure E.2. Kinjiro settled in this house which served as a headquarter of recovery management for administration, police, justice and tax collection. He worked so hard here. He patrolled Sakuramachi nearly every day watching and communicating with farmers, gave suggestions, gave awards (money or rice) to diligent workers whom villagers voted most, reclaimed wastelands, improved irrigation canals and traffic roads, helped the poor and newcomers, etc. He considered the increase of population was important for increasing labors and gave support for the coming-in (some coming-back) farmers. By such recovery practices, the situation of villages recovered to nearly double the yield of rice by 1827 although the next 1828 was a year of poor harvest and the production dropped by about a half.

Figure E.2 (Top) The portrait of NINOMIYA Kinjiro Sontoku drawn by OKAMOTO Shuki (1807–1862) (Photo: Courtesy of Hotoku Museum). (Bottom) His residence (now Ninomiya Sontoku Shiryokan Museum) in Sakuramachi, Shimotsuke (now Tochigi Prefecture), (Photo taken by TANAKA Shigenobu on 26 Nov 2008). Kinjiro lived here for 26 years from 1822 to 1847 and managed recovery of Sakuramachi, Aoki, Kayouno, Yatabe-Mogi, Karasuyama, Shimodate and other villages.

Kinjiro's strategy of increasing labor force by accepting migration and providing some special care to newcomers was quite natural for the development but not necessarily so to native villagers, which caused some conflict between them. Still worse, the dis-satisfied farmers were supported by Kinjiro's new supervisor TOYOTA Seisaku dispatched from the Odawara domain to Sakuramachi in 1827 (41 years old). He represented a dis-satisfied group in samurai class considering that a farmer became a leader of Sakuramachi and kept samurai officers under strict Bundo. Kinjiro could not solve this problem and he submitted a resignation request to the Odawara lord. It was suspended and not accepted. However, early next year in 1829 (43 years old), he suddenly disappeared without any notice to seclude himself in Mt. Narita Temple for three months including 21 days of fasting. During this time, the Odawara lord decided to dismiss Toyota and made Kinjiro the head of reform again. The villagers joyfully welcomed his return and ever since his recovery operation went smoothly. However, in 1831 (45 years old) when the planned 10-year period was over, he considered his promise to the lord to double the tax was not ready yet and requested 5 years extension. It was accepted and by going through Tenpo famines successfully, the target was achieved and the hands were returned to the UTSU Hannosuke family in 1837 (51 years old).

What Kinjiro became enlightened in Mt. Narita Temple was "Ichien-yugo", that is, all players fused into one meaning that everything is interdependent each other and an outcome is a result of integration of all which includes opponent players as well. In fact, when recovery was successful, not only farmers but also samurai joined his management which was an evidence that his recovery concept was accepted by samurai families, too.

The real success was verified in the Tenpo Famine over 1833–1839, especially in 1833 and 1836. It was about 50 years after the Tenmei Famine and its impacts on decreased population and abandoned wastelands were still left in many villages in northeastern Japan. In those villages, another hit of Tenpo Famine was especially serious. Villages in many domains around Sakuramachi suffered from famine and many people starved to death. Nevertheless, in Sakuramachi, there was no suffering or death. Kinjiro waived annual tax in 1832 (46 years old) to express his thanks to farmers who worked so hard for 10 years, and again in 1833 waived tax as it was an agreement to do so in a poor harvest year. Such arrangements were possible only because he had enough savings in his storage. He provided no-interest loan from the accumulated fund, too. Besides, in 1832 before the famine he had instructed farmers to plant millet and save their harvests. It turned out a genius command and how he got this idea is still unknown. At any rate, this success in Sakuramachi became widely known to many domains in Japan and many requests came for him to apply his recovery concept to their domains.

After the Sakuramachi success

Witnessing the successful recovery of Sakuramachi, a request came from Lord OOKUBO Tadazane to work for the Odawara domain in 1834. In fact, he had wanted Kinjiro to recover the Odawara domain first before Sakuramachi. But as Kinjiro's Bundo policy in the Hattori family had been known to the high vassals of the domain, they opposed it and the lord could not ignore them. So he requested Kinjiro to recover his branch domain Sakuramachi first and hoped its success would persuade his vassals. Unfortunately, however, vassals still strongly opposed Bundo and only accepted to give emergency rice and loan to help farmers suffering from the famine. Kinjiro distributed rice and the interest-free loan from the Hotoku-kin (Repay-virtue fund) contributed by the Odawara domain, Sakuramachi and individual contributors. The repay-virtue fund means that the fund is a collection of repayments from the people who got benefits from others' virtue including nature's and parents'. It helped farmers a lot to survive the Tenpo famine, but without Bundo, he could not recover the domain further. Meanwhile his patron Lord OOKUBO Tadazane passed away in 1837. Although the request was renewed, the situation of "no Bundo nor Suijo" did not change. Still, the Repay-virtue fund was kept in function well, but it eventually gave domain officials an impression that Kinjiro (a peasant) was doing the role of domain administration which was against the feudal regime, and the request was suddenly terminated in 1846 (60 years old).

Before Odawara domain's request, Aoki Village (now in Sakura City, Ibaraki Prefecture), a territory of the Kawazoe family had requested Kinjiro a help as they had heard a good reputation of him. The village was located less than 10 km southeast of Sakuramachi and the devastation was similar. The formal request came in 1831 (45 years old) and Bundo was well accepted. He suggested to utilize their abundant grasses "thatch" to renew roofs and also provided Repay-virtue fund (loan). The main task of Aoki Village was recovery of the Sakura River bank. He used his experience in the Sakawa River flood control when he was in Kayama Village and the engineers trained in Sakuramachi helped. The bank was completed by 1833 and the village recovery by 1838 (52 years old). They worked hard with mutual help like "one village one family". Thus they safely survived the serious Tenpo famine.

There were 10 villages in the territory of the Kawazoe family, of which Kinjiro could help, other than Aoki, only another, Kayouno Village (now Ishioka City, Ibaraki Prefecture) west of Mt. Tsukuba and about 20 km further southeast of Aoki. Kinjiro's prescription was to improve infrastructure including irrigation and drainage canals, roof renewal, farming equipment renewal etc. In this case also, Bundo and Suijo worked well and were appreciated.

Kinjiro engaged in recovery of other places too such as domains of Yatabe-Mogi (in Tochigi-Ibaraki Prefectures), Karasuyama (in Tochigi Prefecture), Shimodate (in Ibaraki Prefecture). Both Yatabe-Mogi and Shimodate had huge debt for which severe Bundo was necessary even to return interests. In Shimodate, the necessary salary cut was found to be 28% but the rulers could not accept it. Accordingly, Kinjiro declined his help. But based on his suggestion, a mutual financing system *Shinyu-kou* was established to help samurai officers which worked well and was appreciated. In Yatabe-Mogi, the Bundo was accepted and Kinjiro gave 1000 ryo loan to initiate the development of farming especially in its infrastructure. But eventually the debt increased again and could not continue Kinjiro's plan. In Karasuyama, his philosophy "Ichien-yugo" (all players fused into one) worked well and samurai and farmers literally worked together which was then quite unusual. They saved and invested residuals for agricultural development. They enjoyed the outcome of Ichien-yugo and became prosperous. But officers in Edo who did not know the local reality eventually became dissatisfied in the situation considering their control and superiority were lost, which made them against Bundo.

Works as a Bakushin (Edo government officer)

In 1842 (56 years old), Kinjiro was suddenly summoned by Roju (the top officer of the Edo Government) MIZUNO Tadakuni. Kinjiro halted all the recovery works he was then engaged in and went to Edo. He was assigned to serve as a construction commander of the Tone River and Lake Inba. As he became an officer of the Edo Bakufu (Government) as a Bakushin, he started to use the name "Takanori" (which was read "Sontoku" by others) from this time.

He made a plan, to start from establishing a Repay-virtue fund to improve farmers' life around Lake Inba, and after establishing their support, to build an 18 m wide 18 km long canal to connect Lake Inba to Edo Bay to divert flood discharge from the Tone River. It was a long-term 20-year plan and Roju Mizuno declined it as he wanted an immediate result. In fact, he considered Kinjiro as a construction engineer rather than a social reformer.

Kinjiro requested Mizuno to give an assignment of devastated village recovery. Then in 1844, he was asked to submit a recovery plan of Nikko Shrine territory in Shimotsuke (now Tochigi Prefecture) where the Toshogu Shrine for God Lord TOKUGAWA Ieyasu, the founder of the Edo Government was located. He was so happy on this as he thought it was an official recognition of his recovery method. In 1846 (60 years old) he produced an 84-volume general plan "Nikko Recovery Model" applicable anywhere in Japan. As he was requested a shorter version, he submitted a 60-volume version. It was a product of all members of Kinjiro's school including TOMITA Koukei from Soma domain (in Fukushima Prefecture) who had been with Kinjiro since 1839. Nevertheless, Kinjiro had to wait as long as 7 years till 1853 (67 years old) when a real assignment for Nikko Shrine territory's recovery was given.

In 1847 (61 years old) he moved with his family to Higashigou (now in Moka City, Tochigi) about 5 km north of Sakuramachi where he stayed for 26 years.

Meanwhile, in 1848 (62 years old), the Bakufu-owned (government's) territories, Sougashima Village and later Hanada Village (both now in Chikuzei City, Ibaraki Prefecture) requested for recovery. He gladly applied his Nikko Recovery Model. Sougashima Village had only 10 families with 55 persons in 1843 while one time there were 43 families. He recovered the village, starting from such as roof renewing and toilet repairment, by 3.4 ha reclamation of wasteland, 3.7 km dredging of irrigation and drainage ditches, 2.9 km repairment of access roads, etc. Many people in the neighboring villages volunteered to contribute their labor as well as some for the Repay-virtue fund.

Recovery of Soma domain

Kinjiro had been engaged in, upon request, recovery of Soma Nakamura domain (often simply called Soma domain which is now in Soma and Minami-Soma Cities, Fukushima Prefecture) since 1845 (59 years old). For this, however, he did not go and see Soma but instructed through his top student and a vassal of Soma domain TOMITA Koukei. Tomita as well as all the top officers including Karou KUSANO Masatatsu of the Soma domain were enthusiastic on Kinjio's method (they called it "The Method (Oshiho)") and it became the basis of the Nikko Recovery Model to practice. Among the estimated eventual about 600 applications of his method in Japan, the Soma case was considered the most successful. TOMITA Koukei was the most faithful follower of Kinjiro and he eventually married with Kinjiro's daughter Fumiko in 1852 although she very sadly died the next year due to a stillbirth. Later he wrote books such as "Hotoku-ki" on Kinjiro's life and work after he died (Tomita, 1883).

Soma domain was severely damaged by the Tenmei and Tenpo famines. It used to have a population of 90,000 and produced 170,000 bags of rice but during Tenmei famine in 1783 they decreased to 30,000 persons and 20,301 bags, and in 1816 the debt exceeded 100,000 ryo. During Tenpo famine, 58,680 bags average over 1825–34 decreased to only 4256 bags (7.3%). Population in 1845 was 38,300. Impacts of famines were so serious and still left then. Kinjiro set the Bundo 66,776 bags and started from Narita and Tsubota Villages as they expressed among 226 villages in Soma domain most enthusiastic for reform. Suijo fund or Hotoku (Repay-virtue) fund (residuals from frugal life to be invested for development) contributed for the domain head, samurais and other domain people to get together and realize "all players fused into one". By this Suijo, abandoned fields were renovated, high-interest debts were replaced by interest-free loan, mutually voted persons were awarded, banks were repaired, roofs were renewed, free rice was given to the severest sufferers, etc.

Based on the success of Narita and Tsubota Villages, the application of the method was extended to other villages one by one in the order of enthusiasm and consensus of farmers. Reformed villages reached 50 in 1856 (when Kinjiro died at the age of 70) and 1379 ha wasteland were reclaimed by 1871. When Kinjiro was finally assigned to recover Nikko Shrine territory, the domain head SOMA Mitsutane offered the Edo Government 5000 ryo/year for 10 years, which was of course accepted. It was because Soma domain's debt from Bakurocho loan office of the Edo Government was paid back (because of successful recovery), the interests became no longer necessary to be paid so that this contribution was a token of their thanks to the Government.

Recovery of Nikko Shrine territory and death

In February 1853 (67 years old) Kinjiro finally got an order to start real recovery of Nikko Shrine territory. Nikko Shrine territory had 89 villages, rice production of 20,965 koku, population 21,186, houses 4133, arable field 4216 ha including 942 ha wasteland. It was seriously devastated by Tenmei and Tenpo famines. He set 5 basic actions for recovery based on his "Nikko Recovery Model", namely, to set Bundo from a 10-year average income, to give award money or equipment for diligent farmers and persons dutiful to parents selected by farmers mutual votes, to make a 20-year plan for reclaiming the wasteland, to provide interest-free loans to improve farmers life and to produce straw ropes every day to make sure a self-reliant spirit.

When Kinjiro was about to leave Edo for Nikko in April 1853, he felt sick and delayed a few months. Since then, he had to do the job while fighting a difficult illness. Nevertheless, he went to Nikko and started investigation of villages in July. He found that their irrigation and drainage ditches were the main problem.

Although he did not come back to Higashigou when his beloved daughter Fumi (wife of TOMITA Koukei) died in July 1853 after bearing a dead baby, he had to come back there three months later in October because of his own illness.

Since then, recovery of Nikko recovery was led mainly through his son, NINOMIYA Yataro. The funds he collected were a collection of what Kinjiro had contributed for the Repay-virtue in various places before. He requested them to be returned for Nikko. It was part of Suijo for others although it was not easy. Nevertheless, he could collect enough operational funds and provided no-interest loan to sufferers of high-interest loans, digging irrigation and drainage canals, reclamation of abandoned land, etc.

He stayed in Higashigou for about a year and a half, but when he felt well, he moved to Imaichi, Nikko together with his family in April 1855 (69 years old). Then he patrolled villages in Nikko and instructed his followers. In Todoroku Village, for instance, he was satisfied to see that all villagers understood his concept and worked in the way that all players fused into one. The village was recovered by about 2.8 km of ditches for irrigation and drainage. But because of illness, he had to decline the recovery request of Hakodate domain in Hokkaido. He finally died in October 1856 at the age of 70 years old. But his followers continued the work and his method Hotoku-shiho spread around Japan one after another.

ICHARM is also advocating Hotoku-shiho as the spiritual basis of training disaster management. In its one-year MS program, every year since 2010, students visit "the Sontoku Museum" at the beginning of the semester, and at the end of semester, students vote each other for their most trustable colleague who worked hard and sincerely not only for him/herself but also for the sake of the class and society. ICHARM presents the winner with the certificate of the "Sontoku Award" and a miniature statue of Sontoku as Figure E.3. He/she who got it is supposed to serve as a focal point of the class after graduation for a long time in the future.

Figure E.3 Miniature statue of NINOMIYA Kinjiro Sontoku as part of ICHARM "Sontoku Award". Such statues of a full scale used to be in nearly all elementary schools in Japan as a symbol of hard work and sincerity. Now they are disappearing as considered an old fashion but the spirit behind is still appreciated by many Japanese.

Summary of Hotoku-shiho (Repay-virtue methodology)

Hotoku-shiho, the methodology of recovery that Kinjiro exercised may be summarized as follows:

1. 報徳仕法 (Hotoku-shiho) principles

   (1) 至誠 (Shisei) Act sincerely to others.
   (2) 勤労 (Kinro) Work hard to repay for the virtue of others.
   (3) 分度 (Bundo) Keep living standard within your income.
   (4) 推譲 (Suijo) Invest the residuals for the future of your family and others.
   (5) 積小為大 (Sekisho-idai) Accumulate little efforts to a great.
   (6) 一円融合 (Ichien-yugo) All plyers fused into one. Make every small part work together to achieve an aimed objective.

Hotoku literally means to *repay for the virtue*, that is, people have to work hard to repay for the virtue of others as everyone exists because of virtue of others especially parents, ancestors and the nature. Conversely one's deed decides the fate of future generations.

2. Watch and investigate the real situation of villages.
3. Set a 10-year average yield (income) as the Bundo to live on.
4. Any yield exceeding Bundo level should be saved and used as Suijo, development for the future of yourself and others. Keep being frugal and save for the future sharing with others.
5. Establish a Repay-virtue fund ("Hotoku-kin") made up of contributions from all who saved from frugal life. Lend interest-free loan to pay back high-interest debt to stabilize farmers' daily life.
6. Give a joint responsibility to the community to pay back with one-year extra thanks payment to be contributed to the Hotoku-kin. Five virtues were the basis of this funding system: humanity (仁jin), justice (義gi), courtesy (礼rei), wisdom (知chi), and faith (信shin).
7. Increase labor force by accepting migration from outer areas and supporting them to settle.
8. Dredging irrigation and drainage canals. Roads reforms, houses reforms including roofs and toilets.
9. Working on straw goods such as ropes, shoes, etc. Roofing by thatches.
10. Encourage and give incentives for diligent workers by giving awards who were voted most by farmers themselves.
11. Give support for the widows with small kids.
12. Develop not only rice paddy but also heart paddy for farmers to feel the joy of working for oneself as well as others.

REFERENCES

Fukuzumi, Masae (1933) *Ninomiyaou Yawa*. Iwanami Book Co., Tokyo.

Mokyr, Joel (2020) Great Famine. In: *Encyclopedia Britannica*, 4 February 2020. www.britannica.com/event/Great-Famine-Irish-history (accessed 30 June 2021).

Ninomiya, Yasuhiro (2008) *Life and Thoughts of NINOMIYA Kinjiro*. Reitaku University Publication, Kashiwa, Chiba.

Sumiya, H., K. Inoue, M. Koyama and Y. Tomita (2002) Sedimentary Disasters after the Houei Eruption of Mt. Fuji (1707). *Rekishi Jishin*, 18, 133–147. www.histeq.jp/kaishi_18/21-Sumiya.pdf (accessed 30 June 2021).

Tomita, Koukei (1883) *Hotokuki*. Ryukei Shosha, Co., Nerima, Tokyo.

Uchimura, Kanzo (1908) *Representative Men of Japan*. Keiseisha Book Co., Tokyo.

UNISDR (2009b) *Global Assessment Report on Disaster Risk Reduction 2009 (2009 GAR)*. UN International Strategy for Disaster Reduction. www.preventionweb.net/english/hyogo/gar/report/documents/GAR_Chapter_1_2009_eng.pdf (accessed 26 November 2020).

Wikipedia: 1931 China Floods. https://en.wikipedia.org/wiki/1931_China_floods#cite_note-3 (accessed 30 June 2021).

Wikipedia: Laki. https://en.wikipedia.org/wiki/Laki#:~:text=The%20system%20erupted%20violently%20over,sulfur%20dioxide%20compounds%20that%20contaminated (accessed 30 June 2021).

Wikipedia: Tenmei Famine. https://en.wikipedia.org/wiki/Great_Tenmei_famine (accessed 30 June 2021).

Wikipedia: Tenpo Famine. Tenpō famine—Wikipedia (accessed 30 June 2021).

Index